Our Changing World

Book 1: Threads of Green

*Weaving Sustainability
into the Cultural Fabric*

James Fountain

T R E E L I N E   P U B L I C A T I O N S

ISBNs:

Hardcover: 978-1-963443-00-4
Paperback: 978-1-963443-01-1
eBook: 978-1-963443-02-8
Audiobook: 978-1-963443-03-5

Acknowledgments And Dedication

Before we venture into the heart of this series, let's take a moment to acknowledge the hands and hearts that have been instrumental in bringing this work to life. It is with deep gratitude that I recognize the individuals and communities whose unwavering support and boundless inspiration have laid the cornerstone for these pages. Their contributions have not only enriched this journey but have also been pivotal in shaping the narratives that unfold within.

To start, this series stands as a testament to the guardians of wisdom through the ages—Indigenous peoples, whose very essence is woven into the Earth's rhythms, educators and seekers of knowledge who illuminate our understanding, and you, the valiant souls navigating the sustainability frontier. Your endeavors, though often stretched thin by the magnitude of your obligations and the ceaseless demands for your expertise, are the cornerstone upon which we dare to envision a future brimming with hope and harmony. In these pages lies a shared reservoir of insights, a lighthouse for those teetering on the edge of this crucial movement, aiming to cast light on the path ahead with lessons from my journey, igniting a spark within yours.

The dedication you exhibit, even when overwhelmed and overcommitted, underscores a collective resolve that transcends personal accolades. It's driven by a deep-seated desire for impact, a shared obsession with catalyzing meaningful change in the realms of sustainability, climate action, and human rights. Your commitment speaks to a profound yearning for a deeper connection with the world around us, a connection that is less about individual recognition and more about the collective soul of our planet. In an era shadowed by division, your tireless efforts weave a narrative of unity and hope, showcasing the monumental impact we can achieve through values-driven action and the pursuit of real, tangible solutions. This series seeks not only to honor your undertaking but to amplify it, offering it as a clarion call to all who dream of a tomorrow where our planet is revered, and our collective well-being is cherished.

In this vast journey, Mona, your steadfast support has been my compass, guiding me through the tumultuous and the tranquil alike. Your sacrifices, quietly monumental, from the solitary evenings while I was ensconced in thought or sequestered at the local bar, to the boundless reservoir of patience and encouragement you offered amid my sea of doubts. Our countless brainstorming sessions—naming series, books, and chapters—have been pivotal. This endeavor, imbued with your love and belief, mirrors not just my efforts but the essence of your spirit. The laughter we shared, our nocturnal dialogues, and even the moments of companionable silence have woven a rich tapestry into this work. You have been my muse, my confidante, and my untiring partner through every zenith and nadir. Your love, steadfast and grounding; your incredible life story, a wellspring of inspiration; this milestone bears witness to the enduring strength of our shared journey.

As we stand on the cusp of new adventures, I am buoyed by the knowledge that with you by my side, there are no bounds to the stories we can tell, the worlds we can explore, and the impacts we can make. Here's to the chapters yet unwritten, the tales yet untold, and the journey that continues to unfold with you, Mona, as my guiding star.

And finally, to Liam, my cherished companion whose absence has left a silence too profound for words. I miss you, buddy. In my heart, I envision you in a realm where the forest trails stretch into eternity, where mountain meadows sprawl under the open sky, where tennis balls are abundant, and marrow-filled bovine bones are yours for the taking.

Liam, you exemplified the essence of an ideal Earth citizen—your love, dedication, empathy, and compassion were lessons in themselves, showing me the heights to which we can aspire. Your soul, luminous and guiding, illuminated the path of what life should embody. Through your eyes, I learned to see the world not just as it is but as it could be, filled with love and boundless joy. This book and the series that contains it, while a dedication to the guardians of our planet and the stewards of knowledge, is also a tribute to you, Liam. Your spirit, a guiding star in the vast cosmos of our journey, continues to lead me toward love, empathy, and the profound connections that weave the fabric of life. Thank you for teaching me, for being my compass in exploring what it truly means to live fully and love unconditionally. Your legacy is etched not just in these pages but in my ambition of the being I hope to become.

-JF-

Table of Contents

Preface

Introduction to *Our Changing World*, and Author's Note

As we stand at the crossroads of ecological uncertainty and transformative potential, it is more pertinent than ever that we address the complexities of our changing world with clarity, courage, and deep-seated hope. The journey toward sustainability is a necessary one. The following pages have been crafted to guide us through the uncertainty that shrouds our environmental future. It is with great humility and hope that I present this work as a testament to the potential that lies within our grasp.

The perspicacity with which we must now regard the natural world is unparalleled. We live in times marked by rapid ecological change, and while the challenges this presents are formidable, so too are the opportunities for innovation, growth, and transformation. This book draws upon a chorus of voices —scientists and students, policymakers and professionals, activists and entrepreneurs—all united by a common purpose: to create a sustainable and resilient world for ourselves and future generations.

In weaving this narrative, the intricate patterns that emerge when environmental science, policy, and human values entangle have deeply inspired me. Each thread is vital, and when knit together, they form a tapestry of understanding and insight. I aspire that this work may illuminate the complexities of our time, providing clarity and direction in equal measure.

To effectuate change, one must be armed with the knowledge of both the past and the present. The historical lens through which we will peer opens vistas on lessons learned and paths untaken. We must grasp where we have been in order to chart where we must go. Progressive thought and technological advancements are our allies, but so, too, is the wisdom of traditional practices. This balanced approach will inform our evolving conception of sustainability.

Threads of Green: Weaving Sustainability into the Cultural Fabric is the first book in the *Our Changing World: Navigating Change in a Sustainable, Interconnected World* series. This book is crafted to offer a thorough understanding of fostering resilience and unity amidst the complexities of our constantly evolving world. I have done my very best to present a narrative that is both clear and concise, delving into various dimensions of sustainability and prompting you to contemplate our shared responsibility in creating a more harmonious and sustainable future.

Threads of Green is a compendium of insights and wisdom I have gleaned over many years from the most urban to the most remote communities around the world. I have compiled this collected knowledge to guide you on a journey of exploration and empowerment. I hope to have made the material accessible to all, irrespective of prior knowledge, with straightforward language and concepts designed to provoke deep thought and active participation. You are not just a passive reader; you play a vital role in this dialogue, contributing significantly to the pursuit of resilience and unity.

Amidst the scientific rigor and critical analysis, a narrative of motivation and hope unfolds. Yes, we must confront the stark realities of climate change, biodiversity loss, and resource depletion. But in doing so, we uncover an immense opportunity to reinvent our societies, economies, and individual lives. The sustainable future we aspire to is not just a distant dream but an immediate possibility that demands our dedicated effort.

This series is also a reflection of my own journey—an exploration of the passion that has driven me to spend countless hours researching, writing, traveling to some of the most remote corners of Earth, and advocating for a world where human activity is no longer a threat to Earth's flourishment but a harmonious part of its rhythms. The journey has been both challenging and enriching, filled with moments of stark realization and profound hope. It is this very personal journey that I now share in the hope that it will resonate with your experiences and aspirations.

Welcome to *Our Changing World.*

Embracing Curiosity: A Conversation with Skeptics

In this section, we focus on those who may harbor doubts or questions about the concepts and perspectives shared in this book. Skepticism is welcome as a vital part of any learning journey and is seen as an opportunity for deeper engagement and understanding.

It is important to acknowledge that this book does not aim to provide an exhaustive exploration of a single facet of sustainability. Instead, it presents a rich and varied tapestry of insights, drawing from many cultural perspectives to offer a holistic and comprehensive view of sustainability. This approach is designed to be inclusive, ensuring that the content resonates with a diverse audience, provides a well-rounded understanding of the topic, and appreciates our world's interconnectedness.

To the skeptics among us, we extend an invitation to delve into the pages of this book with an open mind and a curious spirit. The wealth of knowledge and examples from different regions of the globe demonstrate the invaluable contributions of diverse practices and beliefs to our collective understanding of sustainability. By learning from other communities and their unique approaches, we enrich our own perspectives and contribute to a more inclusive and effective sustainability discourse.

This book is crafted to encourage active engagement and critical reflection. Readers are challenged to question their preconceived notions, broaden their perspectives, and embrace the diversity of thought and practice that exists in our global community. This process of active learning and reflection is designed to inspire and motivate, fostering a sense of empowerment and a desire to play an active role in shaping a sustainable future.

Furthermore, this book underscores the significance of community, resilience, and collective action. It highlights each individual's pivotal role in the journey toward sustainability, emphasizing the power of collaboration and mutual respect. For skeptics, this message serves as a reminder of the importance of unity and shared purpose, encouraging a collaborative approach to learning and action.

In conclusion, skepticism serves as a reminder that the journey toward sustainability is not just scientific but profoundly human. It requires institutional transformation and personal- and community-based changes in understanding and behavior. In a world of divergent beliefs and competing priorities, the ability to engage skeptically yet constructively with one another remains one of our greatest challenges—and one of our greatest opportunities.

Series Overview: Exploring Our Changing World and its Central Ideas

Before diving deeper into *Threads of Green,* let's pause to explore the broader context of the *Our Changing World* series, which is a vital part. The subsequent three volumes in this series further unravel the complex tapestry of sustainability, ethics, and business, providing readers with a holistic and nuanced understanding of our challenges and opportunities. The overarching goal of this series is to shed light on the journey toward a more sustainable and just world while also empowering readers to actively participate in shaping our collective future. The aim of these books is to serve as a rich source of information and a wellspring of inspiration. They encourage readers to embrace the power of change and join the movement to create a better world for all.

The *Our Changing World* series is a call to societal introspection and a roadmap toward a sustainable future. It's an amalgamation of ideas and narratives vested in exploring how humanity's impending choices will sculpt the Indigenous environmental legacy. Our journey begins with this book, *Threads of Green: Weaving Sustainability into the Cultural Fabric*, which lays the groundwork by scrutinizing the intricate connections between cultural diversity and unity within the sustainability paradigm. It embodies a firm belief that we can't progress toward a sustainable ecosystem without understanding the kaleidoscope of cultural perceptions and the fused system thinking that global challenges necessitate. From here, the Series will take the following exploration:

A Sustainable Civilization: Unraveling the Threads of Change

A Sustainable Civilization embarks on a comprehensive exploration of the sustainability movement, tracing its roots from the past, navigating through the present, and extending into the future. This journey highlights the essential contributions of diverse cultures, integrating ancient wisdom with contemporary principles to tackle the multifaceted challenges of industrialization, population expansion, and rapid technological innovations. The book lays bare the pressing environmental issues that threaten our planet, emphasizing a pivotal shift from initial conservation efforts to a deeper understanding of ecological interconnectedness.

Central to the narrative are the influential figures and landmark events that have significantly molded the sustainability discourse, offering readers an inspirational glimpse into the evolving ethical landscape that seeks to redefine humanity's bond with nature. *A Sustainable Civilization* delves into the critical battles against climate change and biodiversity loss, underscoring the indispensable role of Indigenous knowledge in weaving a resilient and sustainable future.

By presenting a vision of a world where humanity and nature exist in harmonious coexistence, the book aspires to inspire a collective movement toward sustainable living. It portrays a future where the lessons of the past and the innovations of the present converge to create a sustainable civilization, underlining the importance of unity, innovation, and respect for the natural world in shaping the future we aspire to leave for generations to come.

A Planet in Balance: Exploring the Socio-Cultural Landscape of Sustainability

A Planet in Balance provides a thorough examination of the intricate relationships between sustainability, ethics, and multicultural perspectives. The book investigates the ethical bases that support the pillars of environmental, social, and economic sustainability, presenting a nuanced discussion on how these ethical considerations are essential for understanding and advancing sustainability goals. It explores how individual values, cultural backgrounds, and societal frameworks critically shape our perceptions and engagements with sustainability.

Central to the book's discourse is the vital role that education, art, and storytelling play in enhancing sustainability awareness and motivating societal action. These elements are portrayed as powerful tools for bridging the gap between knowledge and action, effectively engaging diverse audiences and fostering a deeper connection to sustainability issues.

The narrative further extends into the realm of global governance and international cooperation, scrutinizing the mechanisms of policy development and the execution of international environmental agreements. This examination reveals the complexities and challenges of global collaboration, while also highlighting successful strategies that have led to meaningful progress in sustainability efforts.

By providing an in-depth analysis of how ethics, culture, and governance intersect with sustainability, *A Planet in Balance* offers a comprehensive perspective on the socio-cultural dimensions of sustainability. The book encourages readers to reflect on their roles within this global framework and inspires collective efforts toward achieving a sustainable and balanced planet.

The Responsibility Renaissance: Business as a Catalyst for Environmental and Social Ethics

The Responsibility Renaissance marks the culminating insight of the *Our Changing World* series, underlining how businesses are uniquely positioned to drive sustainability and ethical practices within the global marketplace. Advocating a significant paradigm shift, the book urges businesses toward adopting regenerative, waste-minimizing, and resource-efficient practices. It showcases the indispensable role of data and analytics in marrying business operations with environmental sustainability goals through insightful case studies and actionable insights.

Highlighting that sustainability efforts transcend mere ethical obligations to become strategic differentiators, the book offers a variety of strategies for businesses to embed eco-friendly practices across their operations. This encompasses leveraging cutting-edge technologies for sustainable supply chain management and innovating in product and service development.

Leadership's pivotal role in driving the sustainability agenda is underscored, emphasizing visionary leaders' capacity to integrate sustainability with innovation, ethical marketing, and impactful community engagement. The narrative stresses compliance with Environmental, Social, and Governance (ESG) standards as foundational to sustainable business development.

As an essential blueprint for businesses at any stage of their sustainability journey, *The Responsibility Renaissance* delivers a comprehensive framework for weaving environmental and social ethics into the fabric of business practices. Through theoretical exploration and practical examples, it charts a path for businesses striving to be at the forefront of corporate responsibility and sustainability, making it an indispensable resource for leaders and entrepreneurs alike.

The Backstory of the *Our Changing World* Series

What compels a person to delve into the deep river of sustainability, cultural geography, and executive leadership and to emerge with a story to tell? It is the richness of the world's cultural tapestry and the pressing urgency of our environmental crises that fuels the *Our Changing World* series, a narrative canvas I have woven with countless threads of knowledge and experience acquired over twenty years.

My voyage began over two decades ago on the familiar shores of academic rigor, within the bastions of public accounting and professional services firms. Here, the grasp of corporate sustainability strategy was not merely a trade; it was a craft honed through persistence akin to a blacksmith shaping iron. My tenure at prominent Big Four firms was merely the prologue to a much grander tale.

Academia and boardrooms could only teach me so much. It was outside, in the expanse of the living, breathing world, where I would uncover the most poignant lessons. I traversed the globe, journeying through Indigenous territories, sitting with elders, listening to stories of lived experiences from the people living in these remote villages, and understanding the vital pulse of sustainability beat strongest within the communities that have revered the Earth long before the term "sustainability" was even coined.

Amidst these encounters, the inkling that Western ideologies and Indigenous knowledge could converge grew into a bold conviction. As a cultural geographer, I sought to explore not just the physicality of places, but the ethereal sense of "place" embedded in the narratives and rituals that define human-environment interactions.

This exploration transformed me, carving into my consciousness the necessity of a holistic outlook. I saw not only the fragility of nature but the resilience of cultures. Witnessing firsthand the intricate symbioses between communities and their lands, I was heartened by the demonstrated potential for both conservation and sustainable development.

The more communities I visited, the more apparent it became that sustainability is as much about maintaining cultural identity as it is about promoting green technologies. This symphony of sustainability resonates on multiple frequencies—environmental, social, cultural, and spiritual. Each note is crucial for the harmony of the whole.

As I journeyed on, camera in hand and heart open, the diversity of our world unraveled before me in exquisite complexity. Through the lens, each photograph captured a story, a piece of wisdom, a moment where the dance of culture and nature was in perfect step.

My storytelling, thus, is not a mere recapitulation of facts and figures but an endeavor to touch the soul of the reader. The aromas, colors, and textures of distant lands must leap off the pages and draw one into a gentle embrace with the world's myriad societies and ecosystems.

The *Our Changing World* series is inspired by this wealth of discovery. It is a guided expedition that introduces renowned leaders, humble farmers, wise elders, and enterprising youths, all of whom share a common thread—they are the vanguards of sustainability, each in their own right.

In the crafting of this narrative, my objective has never been to merely educate. It has been to awaken a sense of wonderment and responsibility, to ignite the flame of activism within each reader, and to rally a community of kindred spirits that will champion the cause for a sustainable planet.

This is a tale of unity in which science and story coalesce. The scientific underpinnings are undeniably crucial—they provide the framework upon which our understanding of sustainability rests. But the story is the heart, beating life into the framework and inspiring action.

The *Our Changing World* series, therefore, is not a single chorus but a tapestry of voices—a multidisciplinary narrative that seeks to connect the pragmatic solutions of scientific inquiry with the emotional impetus that only a compelling story can provide.

My aim is to capture and translate the soul of sustainability as I have experienced it: in the boardroom and in the wild, in policy and in practice. It is to lay bare the intrinsic connections that interweave humanity with the broader ecology of the Earth.

Thus, each chapter in this series serves as a waypoint on a quest, beckoning readers to journey through the nexus of environment, culture, and leadership. It is an invitation to envision a new paradigm in which humanity thrives not apart from, but as a part of the natural world.

It's a narrative that asserts we are at our best when we recognize the interconnectedness of our world, champion diversity, and walk hand in hand toward sustainability. The *Our Changing World* series is my ode to our planet—a call to all who will listen to join in the dance of change and become stewards of a world that endlessly gives and asks only that we respect its delicate balance.

Introduction:
Weaving a Global Tapestry

I n our world's intricate mosaic, each thread weaves a story, contributes a unique pattern, and reinforces the strength of the whole. As we stand on the precipice of significant environmental, social, and economic challenges, the imperative to weave a coherent global tapestry, celebrating both diversity and unity, has never been more pressing. This book endeavors not just to map the threads of our world's fabric but to inspire, educate, and motivate active engagement in crafting a sustainable and resilient future that reverberates with the echoes of every culture and voice.

To weave such a tapestry requires an understanding that sustainability is not a monochrome concept but a vibrant blend of cultural threads, each infused with its own heritage, ethics, and values. Our task is to honor these diverse perspectives while embracing a shared goal of nurturing our planet. The stories of small villages, the wisdom of Indigenous peoples, and the innovative spirits of urban centers all interlace to form this vibrant vision.

Recognizing the profound impact humans have made upon the Earth marks the beginning of our story. The Anthropocene epoch—an era dominantly influenced by human activity—sets the stage against which our actions must be measured. In realizing that our cultural responses to a changing planet can either harm or heal, we find the impetus to choose wisely and act collectively. We stand at a crossroads, our future shaped by the stories we choose to tell and the actions we choose to take.

Systems thinking offers us a lens through which to view the complexity of interactions within our world. By synthesizing cultural perspectives, underpinned by the understanding that unity can be found in diversity, we begin to see the larger picture. This global vision is essential for anticipating and managing the ramifications of our choices. It allows us to weave a narrative that respects the interconnectedness of all life, understanding that each choice we make sends ripples through the fabric of our collective existence.

Interconnectedness and interdependence, once abstract concepts, are now palpable realities. In this new paradigm, isolationism becomes untenable as the threads of our collective existence become increasingly enmeshed. Embracing these connections facilitates coordinated action, allows for shared resources, and fosters mutually beneficial relationships. Through these connections, we find the strength to address the challenges we face.

Consider the vibrancy that different cultures bring to addressing climate change. Local customs and traditional knowledge may brew a rich source of innovation, coloring our approaches to sustainability. Such diversity can strengthen social resilience, providing lessons that are both enduring and adaptable to places far removed from their origin. The language of culture is potent; it can translate values into concrete actions and shape behaviors with a compelling narrative. By harnessing this power, we begin to align the multitude of human expressions with the rhythms of the natural world.

Indigenous wisdom, juxtaposed against modern challenges, presents a legacy of time-tested practices. It reveals proactive steps to integrate such knowledge for the sake of present and future harmonization with our environment. Religious beliefs and practices have historically guided ethical behavior and can be powerful catalysts for sustainable living. In recognizing the Earth as a sacred trust, faith-based initiatives can compel communities toward responsible stewardship.

Art and music, reflecting the rhythms of our interactions with the Earth, possess unique abilities to motivate change and harmonize conservation efforts with creative expression. These cultural forms enliven the dialogue on sustainability and encourage imaginative solutions. Education serves as a seedbed for these ideas. By embedding multicultural perspectives in the curriculum, we plant the roots for a greener tomorrow and ensure that lifelong learning encompasses the best practices for living in harmony with our planet.

Rural traditions and agricultural methods have long been beacons of sustainability. Through reverence for the land and Ingenious adaptations, these practices represent both a preservation of cultural heritage and an opportunity for innovation and resilience. These cultural insights can infuse green business models with innovation and ethical practices in economics. The interlacing of cultural capital with a sustainable economy holds profound promise for a future where prosperity and planetary health are intertwined.

The social fabric of sustainability is threaded with issues of equity and inclusion. Approaches to health, well-being, and societal balance are varied, yet each becomes an essential strand woven into global sustainability. As diverse cultural patterns inform policymaking, environmental regulations can respect cultural nuances while pursuing global ecological goals. Multilateral cooperation stands as the embroidery that defines responsible planetary governance.

At the heart of this convergence of culture and sustainability is the power of storytelling. Narratives carry the resonant voice of experiences and serve as a lighthouse, guiding us through the fog of challenges that climate disruption presents. In synthesizing these myriad themes, we behold a vision of a world where technology and tradition share a delicate embrace—one inspiring the other, ensuring equitable access to solutions for tomorrow's world.

This introduction lays out the loom upon which we shall weave our story—a narrative rich with color, texture, and pattern. Each subsequent chapter delves deeper into a specific thread, exploring the intricate relationship between sustainability and the cultural tapestry it weaves.

Now, as caretakers of this Earth, join in weaving a tapestry that respects the delicate balance of our ecosystems, honors our cultural diversities, and crafts a resilient and equitable future for all. Let's embark on this journey together, crafting a quilt that is yet unfinished but holds the promise of a masterpiece for generations to come.

Chapter 1:
Cultural Threads in the Fabric of Sustainability

C ultural diversity provides a rich palette of perspectives, insights, and practices in the intricate tapestry that is the quest for a sustainable future. "Cultural Threads in the Fabric of Sustainability," delves into the essence of this diversity, exploring how different societies conceive and enact their relationship with the environment.

Recognition of the pluralistic approaches to environmental stewardship highlights that sustainability is not merely a technical challenge but a complex cultural endeavor. From the spiritual significance ascribed to nature by Indigenous peoples to the sophisticated environmental philosophies in urban centers, no single definition of sustainability transcends cultural boundaries. This chapter brings to light how various cultures contribute unique strands to the overall pattern of sustainability, strengthening the resolve to find common ground on which to build a unified but multifaceted approach to our planetary challenges. By weaving together wisdom from around the globe, we gain a more equitable, effective, and culturally sensitive blueprint for sustainability that honors the knowledge and values of all communities.

Cultural Perspectives on Environmental Stewardship

As we revisit the intricate weave that forms the fabric of sustainability, it's imperative to focus on one of the most vibrant threads: cultural perspectives on environmental stewardship. Culture shapes our relationship with the natural world, influencing everything from the food we consume to the energy we utilize to the rituals we practice. To fully understand and advance sustainability, we must appreciate and integrate the diverse cultural insights from around the world.

The notion of environmental stewardship is not a one-size-fits-all concept. Variations are seen in how different cultures perceive and interface with their environments. For some, it is an inherent duty rooted in traditional beliefs. For others, stewardship through the lens of modern environmentalism. The crucial aspect is recognizing that these varied perspectives can contribute significantly to a holistic approach to sustainable living.

Cultural heritage contains a reservoir of knowledge and practices that have enabled communities to flourish within their ecosystems for generations. By examining these practices, we unveil sustainable pathways, some millennia-old and others adapted to contemporary challenges. This intergenerational wisdom is a testament to the sustainability embedded in many traditional ways of life.

Indigenous communities often embody environmental stewardship through their symbiotic relationship with the land. They hold an understanding of the interconnectedness of life that is vital to sustainability. Acknowledging their rights and knowledge is paramount as we move toward more inclusive sustainable solutions. Integrating these insights into broader environmental discourse enriches our collective understanding and fosters more resilient approaches to ecological challenges.

Environmental stewardship is also deeply influenced by spiritual and religious beliefs. Many faiths preach respect for creation, endorsing an ethic of care for the environment. Religious teachings' moral and ethical dimensions can be powerful motivators for action, galvanizing communities toward stewardship and conservation.

Acknowledging the role of tangible and intangible cultural heritage in stewardship practices, it's evident that traditional agricultural methods, water management techniques, and conservation strategies often enhance local biodiversity and resilience. By valuing and applying such knowledge, we can develop sustainable solutions that are context-specific and respectful of the cultural origins.

One must also consider the dynamic nature of culture and how globalization has affected traditional environmental stewardship. As cultures blend and evolve, so too do ideas of what it means to be an environmental steward. The challenge lies in fostering this evolution while preserving core ecological principles that have stood the test of time.

The interchange of cultural perspectives on sustainability can lead to innovation and a richer set of practices. The notion of adopting agroforestry from the tropics in temperate climates or using water harvesting techniques from arid regions to manage drought exemplifies this potential for cross-cultural learning.

In the pursuit of sustainability, we must also confront the realities of economic development and pressures on natural resources. Cultural perspectives can inform the balance between development and conservation, guiding decisions that consider not only ecological outcomes but also community well-being and cultural integrity.

Diverse cultural perspectives offer a range of strategies for engaging with and educating about sustainability. Storytelling, art, and traditional ceremonies can draw attention to environmental issues in a deeply resonant manner, inspiring action rooted in cultural identity and pride.

Further, youth engagement is crucial in sustainability, and cultural contexts shape how younger generations understand and participate in environmental stewardship. Encouraging youth to engage with their cultural heritage can reinforce the stewardship ethic and ensure its continuation.

As we elaborate on the necessity of cultural perspectives in environmental stewardship, it is also essential to address the risks faced by cultural knowledge systems. Globalization, climate change, and socioeconomic transformations threaten the transmission of traditional knowledge, highlighting the urgency of its preservation and integration into sustainability science.

Finally, we must recognize that the challenge of sustainability is not just a scientific or economic one but a cultural one as well. By engaging with the full spectrum of cultural perspectives on environmental stewardship, we demonstrate respect for cultural diversity and foster a more inclusive and effective sustainable future. Integrating these perspectives requires collaboration, respect, and open dialogue in order to lay the groundwork for a truly global approach to environmental stewardship.

In conclusion, weaving cultural perspectives into the fabric of sustainability enriches our collective understanding and effectiveness in addressing environmental challenges. It offers a diverse array of tools, inspires innovation, and ultimately creates a more inclusive and resilient approach to stewardship. As stewards of the Earth, it's our shared responsibility to recognize, cherish, and learn from the myriad cultural ways of interacting with and caring for our planet.

Defining Sustainability Across Societies

In our journey through the mosaic of global cultures, we've observed a myriad of interpretations and implementations of environmental stewardship. This rich diversity begs a pivotal question: How do we define sustainability across societies? Understanding sustainability requires a multidimensional approach, taking into account its three universally accepted pillars: environment, society, and economy. These pillars are intertwined and critically depend on one another for the health and longevity of communities worldwide.

The environmental pillar is perhaps the most widely recognized aspect of sustainability. It focuses on the conservation and restoration of natural ecosystems and biodiversity. This involves protecting natural habitats, managing resources responsibly, and reducing pollution and waste. However, it's not solely about the intrinsic value of nature; it's also acknowledging that healthy ecosystems underpin human survival and prosperity.

The social pillar of sustainability seeks to ensure social equity, cohesion, and wellbeing. It's centered on the idea that a sustainable society is one that provides its citizens with quality education, access to healthcare, equal opportunities, and a robust framework to support a thriving cultural life. It's in this space that sustainability becomes bespoke to each society, crafted and woven into the societal fabric as each culture sees fit.

Lastly, the economic pillar focuses on creating systems that can support continuous improvement in quality of life without degrading the environmental or social pillars. This means economic growth that is inclusive and equitable, offering fortunes not just for the present, but also safeguarding wealth and resources for future generations.

Acknowledging the shared goals of sustainability around the globe is critical. Despite the diversity in ways societies approach these goals, the common thread lies in the aspiration for long-term viability of our planet and the well-being of its inhabitants. Every society, whether industrialized or Indigenous, faces the challenge of integrating these pillars into a sustainable framework that respects their unique cultural context.

Challenges such as climate change, resource depletion, and social inequity transcend borders and demand a collective response. The solutions require creativity, shared knowledge, and a willingness to learn from one another. It's this shared vulnerability that unites us, galvanizing action across nations and cultures. Together, we confront issues that are far too great for any one culture, government, or sector to tackle alone.

In this shared space, we can understand sustainability not as a strict set of rules, but as a set of principles that guide decision-making processes and development trajectories. While the applications may look different in Copenhagen than they do in Kinshasa, the core objectives remain constant. The principles are adapted and merged into the cultural quilt, respecting local customs, values, and traditions, while keeping the global implications in view.

The beauty of culture is that it can offer novel approaches to sustainability. Indigenous knowledge systems, for instance, provide insights into living in harmony with nature that can complement and enrich scientific understandings. The intersection of traditional practices with modern technology and science opens a fertile ground for innovation in sustainability. Such syntheses can lead to solutions that are both culturally sensitive and environmentally sound.

It's increasingly clear that sustainability cannot be a one-size-fits-all approach. Flexibility and cultural competence are essential, as there is much to be learned from the wisdom embedded in local practices and social norms. This calls for policies and programs that are designed through dialogue with communities and respect the specific ecological, social, and economic contexts within which they operate. Such an approach reinforces community engagement and ensures sustainability initiatives are not just fleeting trends, but that they become integral components of societal values and actions.

Society's engagement with sustainability must also consider intergenerational equity—an ethical dimension that ensures future generations can meet their needs. It's about leaving a legacy that we can be proud of, one that demonstrates not only foresight but also responsibility toward those who will inherit the planet.

At the global level, we're beginning to recognize the implications of living within planetary boundaries. This scientific framework defines a safe operating space for humanity by setting limits on various environmental processes, including climate change and biodiversity loss. These boundaries demand a reassessment of how we coexist with the natural world and implicate cultures in a global dialogue of sustainability.

Empathy and understanding across different cultural contexts are paramount for developing a unified approach to sustainability. This involves a willingness to share knowledge and a commitment to listen as well as the humility to recognize no single culture has all the answers. Sustainability, in this sense, becomes a vast collaborative project, spanning disciplines, borders, and belief systems.

This collaboration also highlights the need for a common language or set of metrics to track progress toward sustainability goals. While qualitative cultural differences are vital, quantifiable measures enable societies to set benchmarks, monitor advancement, and make necessary adjustments. This dual approach balances the tangible with the intangible, the measurable with the meaningful.

As we define sustainability across societies, we open ourselves to the rich tapestry of collective human wisdom and action. By exploring cultural threads and integrating them into the larger fabric, we create a more resilient, adaptable, and compassionate world. It's a world that not only survives but thrives, while celebrating the very diversity that makes our global society so remarkable.

Defining sustainability is an ongoing process, one that evolves as societies change and priorities shift. Yet, amidst this evolution, the commitment to a sustainable future remains unwavering. It is a peregrination defined by cooperation, respect for diversity, and a shared destination—ensuring the well-being of all life on Earth.

Chapter 2:
The Anthropocene and Our Shared Future

I n recognizing the dawning of the Anthropocene, an epoch marked predominantly by human activity shaping Earth's geology and ecosystems, we stand at a crossroads between relentless development and the preservation of our planet. We've dramatically altered landscapes, infused the atmosphere with carbon, and instigated climate change, thereby knitting a complex web of ecological challenges. Yet, within this web lies our shared potential to forge a harmonious future if we recalibrate our relationship with nature.

This chapter delves into that intricate tapestry, acknowledging our profound impact while seeking paths toward a sustainable coexistence. We explore the significance of this new epoch as a call to action, a catalyst for cultural innovation, and an opportunity for global solidarity. Pulling from diverse ideologies and cross-disciplinary studies, this chapter offers a narrative that weaves scientific insights with the powerful stories of our communities, illustrating not only the gravity of our current plight but also the resilient spirit of collective human ingenuity. As depicted here, our shared future reaffirms humanity's capacity for adaptability and the urgent need for a unified approach to embrace the environmental stewardship that secures the well-being of all life on Earth, now and for posterity.

The Epoch of Human Impact

The Anthropocene—a term steadily gaining acceptance—signifies more than a new geological epoch. It heralds an era of profound transformation, within which human activity has dominated the climate and the environment. Our industrial, agricultural, and technological endeavors have etched a lasting signature upon the Earth, characterized by altered biogeochemical cycles, widespread species extinction, and climate change.

This epoch is unique because it represents a period where the collective might of human activity rivals the great forces of nature that have previously sculpted our planet's landscapes. From the Industrial Revolution to the rise of the digital age, our species has remade the world in an image that mirrors our aspirations and our appetites. Yet, this has often come at a steep ecological cost that now demands our attention and action.

For centuries, our species flourished, multiplying and migrating, spreading across the vast expanse of continents, harnessing the power of fire and tools such as wheels, levers, and lathes. And as our numbers grew, so too did our capacity to alter our surroundings. We transformed forests into farmlands, dammed rivers to create reservoirs, quarried mountains for minerals, and built cities that touch the skies. This expansive reach of human endeavor is nothing short of remarkable, but it comes hand-in-hand with profound environmental changes that are now in urgent need of mitigation.

The markers of the Anthropocene are many: altered atmospheric composition due to the combustion of fossil fuels, unprecedented rates of land use change, the proliferation of plastics in the oceans, and the mass extinction of species driven by habitat loss and climate change. Each of these markers tells a story of innovation and enterprise, but also of imbalance and oversight.

Climate change, perhaps the most spoken-of consequence of this epoch, tests our societies' limits and moral responsibilities. Rising sea levels, extreme weather events, and shifting agricultural zones present not only logistical challenges, but also questions of justice, as those least responsible for emissions often face the most severe impacts.

However, within this epoch lies the potential for awakening. The very understanding that our actions have been the architect of such profound changes can empower humanity to become the stewards of a more balanced and sustainable Earth. This awareness invites a renaissance of conservation, innovation, and respect for the natural systems that sustain us.

Indeed, as agents of change, humans can rectify past harms. Though technologies and infrastructures have been sources of disruption, they also possess the potential to rebuild, restore, and rejuvenate. Renewable energy sources such as wind, solar, and hydroelectric can reduce our carbon footprint. Reforestation and regenerative agriculture can sequester carbon and enhance biodiversity.

Our impact is not confined to environmental dimensions; the Anthropocene encompasses social and cultural ramifications. Such a multi-faceted epoch necessitates discussions about environmental issues integrating perspectives from economics, sociology, history, and beyond. As we steer toward sustainability, the solutions we seek must be as intersectional as the challenges we face.

The emergence of circular economies, which emphasize reuse and recycling over the linear model of consumption and disposal, showcases our capacity for adaptive change. In a circular economy, waste is minimized and products are designed for longevity, repairability, and afterlife utility. Such transformation in economic paradigms can significantly diminish human impact and foster a resilient, sustainable future.

An imperative part of shaping the Anthropocene toward sustainability is acknowledging the disparity in impacts and responsibilities across different populations. Equity must be at the heart of the transition, along with efforts to ensure that the benefits of sustainable practices and technologies are shared by all. This inclusivity is not just ethical, it is pragmatic, as it engenders broader support and participation in the endeavors for a sustainable world.

The Anthropocene calls for an evaluation of how we perceive progress and success. It asks that we measure wealth not solely in GDP, but also in the health of our ecosystems, the resilience of our communities, and the equity of our societies. Sustainability should not be an afterthought, it should be the fundamental criterion guiding development.

The need for education and awareness is paramount. By understanding the intricacies of the challenges we face, individuals and communities can be motivated to act. Education should thus strive to imbue all citizens with the knowledge and values critical for the stewardship of our planet.

There is an artistic and creative dimension to our response to the Anthropocene: the ability to envision and manifest a world that marries human flourishing with ecological integrity. This endeavor is as much a cultural project as it is a scientific or political one, requiring the imagination to conceive alternative futures where humanity thrives in symbiosis with nature.

In conclusion, the Epoch of Human Impact is a clear call to action. It is an invitation to reexamine our relationship with the world and forge a sustainable, equitable, and just path. The Anthropocene is not an end but a beginning. It's a foundation upon which a legacy of conscious caretaking and respect for the planet can be built. By embracing this call, we craft not only a narrative of survival but one of thriving for all species sharing this Earth.

Cultural Responses to a Changing Planet

As the narrative of the Anthropocene unfolds, it is not just the physical aspect of our planet that is transforming. The cultural responses to these ecological shifts are equally crucial in guiding humanity's path toward a shared, sustainable future. Cultures around the globe are reevaluating and adapting their traditions and practices in response to the burgeoning reality of climate change and environmental degradation.

For centuries, our diverse cultures have provided a rich tapestry of wisdom and practices, teaching us myriad ways to live in harmony with nature. Today, in the face of unparalleled environmental challenges, these cultural reservoirs are being tapped into as never before. Some communities turn to ancestral knowledge for insight into sustainable living, while others forge new cultural practices that align with the ecological imperatives of our times.

Art has always been a reflection of the times and, in today's context, it has become a powerful medium to express ecological concerns and inspire change. Globally, artists are harnessing their creativity to provoke dialogue and bring environmental issues to the forefront of public consciousness. Murals depicting threatened species, sculptures forged from reclaimed materials, and music that evokes the beauty and plight of natural landscapes all serve to unite communities around a common concern for our planet.

In literature and film, stories and narratives are evolving to include themes that question humans' relationship with the environment. Through these stories, authors and filmmakers are charting new ways of thinking, understanding, and relating to the natural world. We see characters and communities wrestling with the impacts of climate change, resource scarcity, and environmental injustices, mirroring real-world conflicts and offering insights into potential solutions.

Language, too, is evolving. New words and concepts—such as "ecocide," "anthropause," and "solastalgia"—have entered our lexicon as we strive to articulate the intricacies of our changing reality. Language shapes thought, and as our vocabulary expands to include these terms, so too does our collective awareness and response to the ecological challenges we face.

In the realm of education, curricula worldwide are integrating sustainability concepts, ensuring the next generation is equipped with the knowledge and values necessary to live in a more sustainable world. Environmental ethics are being taught alongside traditional subjects, aiming to embed a consciousness for sustainability in the very fabric of learning.

Religious and spiritual communities are also playing a role in shaping cultural responses. Many are reinterpreting their doctrines in the light of current ecological crises, calling on adherents to act as stewards of the Earth. This leads to faith-based initiatives promoting conservation and sustainable living practices, infusing ancient wisdom with modern conservation efforts.

Food culture is adapting to the needs of a strained planet. There is a growing movement toward plant-based diets, reduced food waste, and support for local and sustainable agriculture. These shifts in dietary practices are not just about personal health, they signify a broader recognition of the interconnectedness between how we eat and the health of our environment.

Indigenous populations are a vanguard of cultural resilience. They offer profound insights into sustainable living based on centuries of close-knit relationships with their environment. Efforts to preserve and integrate Indigenous knowledge into wider society are gaining momentum, as these cultural practices offer tangible benefits to biodiversity conservation and climate change mitigation.

Cultural responses are unique and differ vastly across societies and communities. What is common, however, is the growing realization that our actions must be collectively redirected toward sustainability. Every cultural transformation, whether it is a shift in consumption patterns, new artistic movements, or the re-emerging relevance of traditional practices, contributes to a larger, global effort to address the planet's changing needs.

Businesses and corporations are recognizing the value of culture in sustainability. This recognition is birthing new business models that incorporate cultural heritage and practices into sustainable enterprise. From multinational corporations to local startups, there is a burgeoning understanding that cultural sensitivity can drive innovation and resonate with environmentally conscious consumers.

At the policymaking level, there is an increasing acknowledgment that legal frameworks and regulations must account for cultural diversity. Environmental policies that are culturally informed are not only more equitable, but also more effective. They harness the full spectrum of human creativity and insight, ensuring legislation supports rather than hinders sustainable cultural practices.

On a community level, the revitalization of shared spaces—parks, community gardens, and cooperative hubs—serves to cultivate a collective identity around green living. These are spaces where cultural integration blooms and sustainable practices take root, strengthening community ties and bolstering the local response to environmental challenges.

Finally, in looking ahead, envisioning future societies that are both sustainable and culturally vibrant is becoming a necessity. Imagining utopias gives concrete form to our aspirations, serving as guiding stars for the cultural evolution needed to address the looming threats of the Anthropocene. The cultural responses we foster today will be the foundations upon which these future societies are built.

Our planet is changing and, with it, our cultures. The Anthropocene demands a cultural metamorphosis; one that aligns human creativity, wisdom, and innovation with the needs of our environment. As stewards of the Earth, it is our collective cultural heritage, shaped and reimagined, that will determine how gracefully we navigate the tide of environmental change sweeping across our planet.

Chapter 3:
Systems Thinking and Cultural Synthesis

A s we delve into the rich tapestry of human societies and their bonds with the natural world, we come upon a pivotal mode of thought: systems thinking. This approach is an analytical framework for understanding the dynamic and complex interplay of elements within and outside a given system. It's defined as the ability to recognize patterns and relationships, to see both the components and the whole, and to anticipate the indirect consequences of actions.

This chapter explores how systems thinking acts as the bedrock for cultural synthesis, a concept essential for fostering sustainability in a world brimming with diversity. We delve into how embracing complexity brings to light the multifaceted interactions between human activities and environmental vitality. Here, we unite the myriad of cultural nuances and sustainability practices into a concatenated systems approach, emphasizing that harmonizing these can engender resilience against the throes of global challenges. Empowered by this outlook, readers are encouraged to think in systems—not in isolation—recognizing that every action has a ripple effect, affecting the socio-ecological balance in unpredictable ways. This chapter invites you to reimagine how cultures around the globe can synergize to create a web of sustainability that is as intricate and resilient as the ecosystems we aim to preserve.

Embracing Complexity in Sustainability

In our journey through the tapestry of sustainability, we have waded through the waters of cultural perspectives, grappled with the weight of the Anthropocene, and now stand before the inevitable crossroads of complexity and simplification. This section, dedicated to embracing complexity in sustainability, delves into the crux of systems thinking and cultural synthesis. It's here where we unfold the nuanced interplay between environment and culture and the critical importance of embracing the layered reality of sustainable practices.

At the heart of sustainability lies the understanding that our world is a woven web of causes and effects, where simple actions ripple into complex outcomes. The concept of sustainability, in its essence, demands intricate balance between human needs and environmental limits, between the local and the global, and between immediate actions and long-term consequences. To embrace complexity is not to be daunted by it, but to recognize the multifaceted interactions that define our world.

Modern society often strives for clear-cut solutions and binary narratives. However, true sustainability requires us to move beyond this reductionist thinking. As we consider ecosystem services, for instance, we find they are not merely commodities to be traded or managed, but are deeply intertwined with cultural identity and societal well-being. The water that flows through a river is not only H_2O, it's a life source, a cultural highway, a character in myth, and a measure of ecological health.

Recognizing complexity, therefore, involves acknowledging the limitations of a one-size-fits-all approach. Bioregionalism is a testament to this, as it encourages us to adapt our methods and define sustainability within the unique context of each specific region by incorporating local knowledge and conditions. This philosophy champions the localization of our actions, albeit within the broader understanding of a globally interconnected system.

Diversity—both biological and cultural—is a wellspring of resilience. When we value and maintain diversity, we enable systems to absorb disturbances and adapt to change. This diversity provides a vast repository of knowledge and strategies for surviving, and even thriving, in the face of adversity. From farming techniques that vary with the landscape to diverse diets that make communities less vulnerable to single crop failures, complexity and variability are assets, not challenges to overcome.

Acknowledging complexity also means embracing the unexpected. In socio-ecological systems, there is a phenomenon known as emergence, in which new properties and behaviors manifest that were not predictable from the system's components alone. This can lead to innovation, adaptation, and new challenges and conflicts that require dynamic and flexible responses.

The governance of sustainable systems is, by necessity, a complex affair. It involves the delicate balance of empowering local autonomy while ensuring that the collective impact on the planet is within safe boundaries. Polycentric governance, where multiple decision-making centers operate at different scales, offers a model that can navigate this balance, fostering cooperation and learning across diverse contexts.

In embracing complexity, we must recognize the real-world applications, such as urban planning and infrastructure development. Smart cities, for example, incorporate technology not merely for efficiency, but to create responsive, adaptable, and resilient systems. They marry data with the lived human experience of the city, valuing quality of life alongside economic and environmental outcomes.

Sustainability science itself is an inherently interdisciplinary field. It unites biology, ecology, engineering, economics, and social sciences, among others. This integrative approach is paramount for dealing with complex sustainability challenges where isolated knowledge simply can't suffice. By drawing from multiple disciplines, solutions can emerge that are robust, equitable, and deeply rooted in a holistic understanding of the world.

Understanding complexity also involves recognizing the role of time. Temporal dynamics present challenges, such as the lag effects of pollution or climate change where actions and their consequences are mismatched in time. A sustainable outlook acknowledges these temporal complexities and strives for foresight and taking actions today that will benefit generations far into the future.

Engaging with complexity means moving beyond the technocratic mindset that often pervades sustainability conversations. While technology offers phenomenal tools, the solutions we seek equally reside within the social and cultural layers of society. Social innovation is just as crucial as technological innovation when it comes to fostering sustainable lifestyles.

Culture—in all its colorful forms—is a lens through which we interpret the world and our place in it. In embracing complexity, we must also embrace cultural narratives that shape our values and behaviors. These narratives can foster a deep connection with nature and a sense of stewardship or, conversely, drive consumption and detachment.

Education systems and pedagogies, too, can either compartmentalize knowledge or teach us to think in systems. When we train our minds to perceive relationships and patterns and to think critically and creatively, we lay the groundwork for sustainability that acknowledges the world's complexity.

In conclusion, the path to sustainability is not a straight, narrow one, but a labyrinth rich with learning and life-sustaining complexity. To walk this path, we must be willing to entertain ideas that challenge our worldviews, engage in meaningful dialogue with those who hold different perspectives, and adapt with both humility and determination when faced with the planet's intricate realities.

With these thoughts as our guide, we can begin to unwrap the next layer: unity through diversity in the systems approach, in which the beauty of complexity truly shines through as a key to creating a sustainable and resilient world for all.

Unity Through Diversity: The Systems Approach

At the heart of systems thinking lies the profound understanding that diversity and unity are not opposing forces, but rather complementary elements that enhance the resilience and adaptability of any given system. The systems approach recognizes that cultural diversity, much like biodiversity, strengthens ecosystems by bringing a multitude of perspectives and practices that can be synthesized for effective problem-solving. It's here where sustainability transitions from an abstract concept to a tangible outcome of diverse but interconnected actions and attitudes.

Drawing from the numerous arrases of global cultures, systems thinking encourages the integration of knowledge, traditions, and innovations. Each culture provides a unique lens through which environmental challenges can be viewed and addressed. Within this rich mosaic, individual cultural practices contribute to the collective wisdom necessary for forging a sustainable future. Acknowledging varied cultural insights underpins a broader, more inclusive approach to sustainability.

The systems approach does not suggest homogenizing cultural practices, but rather understanding their underlying commonalities. For instance, while farming techniques may vary drastically from the terraces of East Asia to the rainforests of South America, at their core, these practices are aligned to work with local ecosystems rather than dominate them. This is the shared principle that systems thinking brings to light, fostering a unity of purpose amidst a diversity of practices.

By embracing cultural differences and leveraging them for shared goals, we can tackle the complex socio-ecological issues that define the Anthropocene. We are facing problems that are not only technical, but also cultural and moral in nature. Systems thinking, therefore, calls for transdisciplinary collaboration that transcends narrow scholastic boundaries and incorporates the lived experience and traditional knowledge of varied communities.

Moreover, systems thinking recognizes the nonlinear nature of sustainability challenges. Simple cause-and-effect relationships are rare in complex systems where multiple factors influence each other in dynamic and sometimes unpredictable ways. Cultural influences can be diversifying factors that introduce new variables and potential solutions into the dynamics of the system, providing a robust and flexible framework for addressing these intricacies.

Unity through diversity also requires an equitable platform where voices from different socioeconomic backgrounds are heard. In the spirit of systems thinking, these voices' synergies can guide communal efforts toward sustainable livelihoods. Listening to marginal voices, which often carry generational wisdom about living in harmony with nature, enriches the conversation and reinforces the social fabric supporting sustainability efforts.

Systems thinking encourages intersectional analysis and engagement. This approach understands issues like climate change cannot be separated from social justice concerns and solutions must accommodate the varied needs and contributions of different cultural groups. Sustainable development, therefore, necessitates a systems approach that is both inclusive and integrative, and recognizes diverse cultural practices and understandings can collectively shape a way forward for humanity.

One of the richest manifestations of this unity through diversity is seen in the realm of ecological knowledge. Indigenous and local environmental knowledge systems are not just archives of past practices, but living libraries of adaptability and resilience. They demonstrate the capability of human cultures to evolve with changing conditions, often in sync with natural cycles and resources. Systems thinking thus advocates for a conscious synthesis of modern scientific methodologies with the time-tested practices embedded within these rich cultural narratives.

This synthesis is not just a theoretical idea, but a practical resilience-building strategy. The diverse knowledge systems can offer a portfolio of options for communities seeking to adapt to local environmental changes. These options are vital in an era when climate change presents not one but multiple location-specific challenges that require tailored solutions.

Systems thinking also underscores the importance of scale in sustainability efforts. Diversity entails scaling solutions from local to global, acknowledging that local actions have implications at the planetary level. When patterned into the larger system design, local cultural practices can contribute to building adaptive capacity on a broader scale. Through a systems lens, aggregating numerous local success stories leads to global impact.

Another facet of the systems approach is the feedback mechanism, where cultural practices are not statically perpetuated, but are evaluated and adapted according to changing environmental and social conditions. This dynamic allows for cultural diversity to not merely survive but thrive, informing and being informed by the sustainability efforts in which they partake. It encourages an ongoing conversation between tradition and innovation—a vital exchange for cultures that aspire to be sustainable.

Furthermore, unity through diversity accommodates the proliferation of niche innovations. The systems perspective recognizes the need for a range of strategies to emerge, some of which will be highly context-specific or even counterintuitive from a globalized standpoint. These innovations, grounded in local cultural understandings, can then disseminate throughout the system, sometimes radically shifting the paradigm of sustainability practice.

To operationalize this systems approach, policymakers and practitioners are encouraged to actively pursue participatory engagement. This means facilitating platforms where multiple stakeholders—from government leaders to grassroots activists, scientists to local community knowledge holders—can share insights and cocreate sustainability strategies. This collaborative atmosphere is vital for cultivating the unity in diversity needed to weather the storms of this era and beyond.

In conclusion, the systems approach to unity through diversity is not merely about recognizing differences but weaving them into a coherent strategy for sustainability. It calls for openness to new ideas, respect for traditions, and recognition that the paths to sustainability are as diverse as the cultures that tread them. In doing so, we can build a resilient tapestry that withstands ecological and social turbulences, a fabric strong in its diversity and unbreakable in its unity.

Chapter 4:
Embracing Interconnectedness and Interdependence

I n the contemporary mosaic of global sustainability, acknowledging our world's inherent interconnectedness and interdependence is fundamental. This chapter propels us into the intricate web of relationships that transcends borders, societies, and ecosystems. It urges us to embrace our collective existence and to think of ourselves not as separate entities, but as part of a greater whole. This realization comes with the clear acknowledgment that our existence's environmental, societal, and economic pillars are so tightly woven that the ripple effects of actions in one corner of the globe can reverberate across oceans with profound impacts.

In navigating this complex relationship, we uncover the paradoxical interplay between isolationism and interconnectedness, where the latter can be a potent tool in combatting global challenges, including climate change and social inequality. By leveraging our global connections, we can create a robust framework for sustainable impact that integrates ethical considerations and respects cultural diversity. It's about revamping our narrative, shedding the obsolete skin of isolationism, and adapting to a paradigm of coexistence that fosters mutual progress and resilience. Through a lens of cultural sustainability, readers are invited to explore how a deeper understanding of shared fate and collective capability can pave the way toward lasting change. The ideas in this chapter urge each of us to step into roles as stewards of our communities and the wider world that holds them.

Navigating the Tensions between Isolationism and Interconnectedness

In a world where the echoes of isolationism reverberate against the reality of global interdependence, we must navigate these complex tensions with caution and courage. Isolationism, with its seductive allure of national self-sufficiency, often fails to recognize that the environmental, economic, and social challenges we face do not respect man-made borders. Conversely, interconnectedness presents a conundrum; while it might exacerbate issues like over-dependence on global supply chains, it offers invaluable opportunities for collective problem-solving and innovation.

Through a careful balance and recognition that sovereign actions can have far-reaching global repercussions, we can harness the positive aspects of interconnectedness—such as shared knowledge, diversified risk, and collective action—to address sustainability within an interdependent global framework. The urgency of climate change demands no less than a concerted and cooperative endeavor that honors autonomy while embracing the necessity of collaboration for the greater good.

Isolationism vs. Interconnectedness: Sustainability in a Disconnected World

As we navigate from the last theme of unity through diversity, we're ushered into the nuanced discussion of isolationism and interconnectedness. It's imperative to explore how an ostensibly fragmented world can still strive for sustainability. Isolationism manifests through nations or communities adopting insular policies, often at the expense of international cooperation, while interconnectedness is a state of mutual dependence and collaboration among different geopolitical, cultural, and social entities. This dichotomy poses the questions: Can sustainability be achieved in a disconnected world, and to what extent is interconnectedness essential for a sustainable future?

Historically, isolationist stances were presumed to protect resources and traditions. Yet, in the face of mounting environmental pressures, the viability of such a stance comes under scrutiny. Climate change, biodiversity loss, and resource depletion don't confine themselves within political borders, making us question the effectiveness of an insular approach.

Consider the global nature of our ecosystem. Environmental degradation in one region can profoundly affect another. For example, deforestation in the Amazon affects rainfall patterns not only in Brazil, but as far north as the Midwest of the United States. Thus, we must acknowledge that our environmental fates are intertwined in a manner that isolationism fails to address effectively.

The benefits of interconnectedness for sustainability are manifold. One significant advantage is the shared pool of knowledge and resources. By recognizing the interconnected nature of our current challenges, solutions through technological exchange and collaborative research initiatives have greater potential to emerge.

Interconnectedness also brings into focus the global value chains that shape economic, social, and environmental outcomes. For example, green supply chain management initiatives can drive significant sustainability improvements by traversing national borders and instilling environmentally friendly practices across industries.

But, interconnectedness isn't without challenges. It opens doors to exploitation, inequity, and cultural homogenization. The adverse effects of global interconnectedness, such as the rapid spread of disease as evidenced by the COVID-19 pandemic or the economic disparities exacerbated by globalization, need to be critically evaluated and addressed to harness its potential for sustainability.

Communities that decide to lean toward isolationism may do so to protect local ecosystems or to preserve cultural uniqueness, but they do not exist in a vacuum. The environmental damage incurred far away can still upset the delicate ecological balance of remote and isolated communities. Even in disconnection, the world's ecological system remains an interconnected web.

To thrive sustainably in this seemingly disconnected world, there must be a balance between isolationism and interconnectedness. Local self-reliance in food, energy, and culture can complement global environmental initiatives, creating a resilient patchwork of local and global stewardship. To this end, policies must enforce that while economies operate globally, they still respect and nurture smaller, community-based systems.

With these ideas in mind, policymakers need to understand that isolationism in environmental terms is a misnomer. While respecting the autonomy and decision-making of all cultural entities, there needs to be an overarching framework that ensures the environment—which knows no border—is treated as a shared treasure. Global policies and agreements like the Paris Agreement serve as conduits to this purpose, fostering interconnectedness while allowing local autonomy.

Economically, the shift toward interconnectedness can aid in sustainable development by sharing the burden and benefits. Funding mechanisms and technology transfers from more affluent nations can underpin sustainability efforts in developing regions, showcasing how interconnectedness can bridge the sustainability gap between different socioeconomic realms.

Furthermore, interconnectedness can foster cultural sustainability. Sharing struggles and successes can intertwine cultural narratives, breeding empathy and shared responsibility. Programs and partnerships encouraging cultural exchange and education can serve as the foundation for a more cosmopolitan stewardship of our planet.

We must also recognize that interconnectedness in a disconnected world is not just about tangible resources and policies. It's equally about intangible values, such as trust, empathy, and mutual understanding. These are the underpinnings of a sustainable world. They foster cooperation that transcends mere transactional relationships.

The juxtaposition of isolationism and interconnectedness within sustainability demands a nuanced approach. While recognizing the right of communities to safeguard their resources and identities, there must be an acknowledgment that in an interconnected environmental system, collaborative action is not just beneficial—it's imperative. Therefore, the transition toward sustainability in a disconnected world lies in celebrating diversity while embracing the unifying thread of our shared ecosystem.

Impacts of Isolationism: Environment, Society, and Economy

The choice to adopt isolationist policies, national or community, has significant consequences across multiple facets of global sustainability. Isolationism, often characterized by the insulating policies and practices a society enacts to safeguard its interests, can lead to a fracturing of the rich tapestry that a globally interconnected community offers. In this disconnection, we see multi-pronged impacts that can deleteriously affect the environment, fracture societies, and destabilize economies.

The environmental impacts of isolationism are glaring when nations barricade themselves from collaborative efforts on climate change and biodiversity loss. Efforts become piecemeal and less effective without a shared commitment to combatting global environmental challenges. Pulling away from international agreements like the Paris Climate Agreement can exacerbate environmental decline, as the cumulative effects of individual actions are less than the unified strides humanity could take together.

In social terms, isolationism can breed xenophobia and reduce the exchange of cultural wisdom that strengthens community resilience to changes, including those brought by our evolving climate. The loss of social cohesion has been linked to increased mental health issues and weaker support networks, which are crucial in times of socioeconomic or environmental stress.

On the economic front, isolationist policies often aim to protect domestic markets, yet can lead to trade wars that hurt global and local economies. Scholars argue that such policies can lead to inefficiencies and supply chain disruptions, causing increased consumer prices and making the market more volatile.

Isolationism directly affects environmental policies. For instance, a country refusing to participate in global conservation efforts might choose to unsustainably exploit natural resources. This exploitation can have a domino effect on global ecosystems, exemplifying the interconnectedness of our environmental systems. No single nation can act without influencing the broader ecological equilibrium.

Concerning society, isolationism can cause a withdrawal of multicultural influences that strengthen social fabrics. It ignores the benefits of diverse ideas and practices that have long been exchanged and improved upon throughout human history. Such a withdrawal from global discourse on societal issues restricts the ability to adapt to global shifts, creating an inward-looking populace less prepared for global crises.

From an economic standpoint, protectionist trade policies often result in retaliation by other nations, reducing trade. Reduced trade can have far-reaching effects on national economies and the global economic system, constraining growth and innovation and may ultimately lower the standard of living within protective borders.

Climate change is an inherently global challenge, and isolationist stances only serve to hinder progress. National endeavors to reduce emissions fail to stack up against the potential impact of coordinated global action. The failure to collaborate on climate initiatives can result in a disjointed approach that is both ineffective and significantly more expensive.

Isolationism also tends to create societal division, often fostering an us-versus-them mentality. Such divisions limit empathy and understanding across cultures, which is detrimental to building the societal resilience required to deal with shared challenges such as pandemics, economic crises, and the effects of climate change.

The economic drawbacks of isolationism become pronounced in the realm of innovation. By cutting off the flow of international talent and the vibrancy of cultural exchange, countries reduce their access to new ideas and technologies that drive economic development. A nation that closes its doors to the world may lag in the technological race integral to a flourishing, modern economy.

By removing global barriers and regulations, isolationist environmental policies might initially seem beneficial for certain local industries, such as fossil fuel production or logging. However, the long-term consequences of environmental degradation, including soil erosion, water pollution, and biodiversity loss, can result in dire economic costs and jeopardize future generations' well-being.

Isolationism impacts social welfare systems, too. When a country is disconnected from the global flow of aid and humanitarian outreach, it struggles to cope with large-scale disasters or economic downturns. This challenge is acutely felt by marginalized communities that often rely on international support structures to augment domestic resources.

Economic isolationism can also disrupt agricultural sectors on a global level, a primary example being tariffs and subsidies. Such practices can harm international relationships and lead to inefficiencies within the agricultural sector, eventually impacting food security and pricing at both local and global levels.

An isolated approach to environmental governance can neglect the shared nature of our planet's air and waters. Pollutants and emissions do not respect national borders; thus, an isolated policy framework is inconsistent with the nature of environmental systems. It is ineffective at best and harmful at worst.

Social isolationism can be particularly damaging during global health crises. The COVID-19 pandemic underscored the need for global solidarity and information sharing. Societies that shunned international cooperation often found themselves ill-prepared and therefore faced dire public health and social consequences. Achieving a robust societal response to global challenges requires a collective effort that transcends isolationist tendencies.

Fiscally, the digital age and the growth of global markets mean economies are more intertwined than ever before. Isolationist economic policies stifle potential growth by ignoring this interconnected marketplace. They can also impede the development of emerging markets, which can be prime drivers of global economic stability and growth.

In summary, the impacts of isolationism on the environment, society, and economy are complex and intertwined. Each instance of withdrawal on the global stage has echoes that reverberate far beyond a single nation's borders. Sustainable progress, in contrast, is fostered through cooperation, mutual understanding, and shared responsibility. It is incumbent upon us as a global community to champion policies of interconnectedness where growth and sustainability can flourish not in isolation but in harmony with the world.

Leveraging Global Connections for Sustainable Impact

This era of profound interdependence compels us to leverage global connections meaningfully to create sustainable impact. As we stand at the crossroads, the power of collective action to address environmental crises has never been more evident. Fostering sustainable impact requires harmonizing local efforts with international expertise and resources, creating a synergy where ideas, technologies, and strategies for resilience are shared openly and adapted locally.

Through global collaborations, communities can implement innovative practices such as decentralized renewable energy systems, which have been successful in different countries and allow for the development of a less carbon-intensive and more equitable economic system. These collaborations not only cultivate practical benefits, but also reinforce a global culture of sustainability, shaping a world where the ripple effects of positive change in one region can trigger waves of progress in another.

The essence of sustainable impact lies in understanding the health of our planet and communities doesn't recognize borders; therefore, our actions and strategies for preservation and growth must transcend them as well. Sustained dialogue among nations, continuous knowledge exchange between academics and practitioners, and an unwavering commitment to mutual progress form the bedrock of a resilient and vibrant Earth community.

Harnessing Interconnectedness to Overcome Global Divides

In the zeitgeist of global environmentalism, isolationism presents itself as an insidious thread capable of unraveling the fabric of collective progress. However, the power of interconnectedness, like the roots of an ancient forest, holds profound potential to bridge the divides that challenge our pursuit of sustainability. By leveraging these connections, which span across technological, ecological, and cultural realms, society can foster a synergy that propels us toward a cohesive global community.

Recognizing our interwoven existence begins with appreciating the myriad threads that create our world's tapestry. In this interconnected matrix, every action has ripples that reach far and wide—economically, environmentally, and socially. Acknowledging this intricate web is crucial in comprehending how local actions can have global consequences and vice versa. This systemic view becomes the bedrock upon which strategies for overcoming divides can be structured.

Interconnectedness requires robust communication channels that traverse geographical and cultural barriers. In today's age, the digital realm offers unparalleled opportunities for cross-pollination of ideas, fostering what can be seen as a digital ecosystem that mirrors our planet's biodiversity. When used thoughtfully, digital networks can propagate a shared consciousness of sustainability challenges and incubate solutions enriched by their contributors' diversity.

Transcending differences to address climate change demands more than simply recognizing common ground—it necessitates active collaboration. Collaboration across borders allows for the pooling of resources, knowledge, and expertise, ensuring that when one region pioneers a successful sustainability initiative, it can be adapted and implemented elsewhere. This mirrors ecological mutualism, where different species work together for mutual benefit, a concept that can and should be mirrored in human cooperation to solve global issues.

Education plays a pivotal role in cultivating a culture of interconnectedness. Through inclusive curriculums that celebrate cultural diversity and emphasize our shared responsibilities to the planet, future generations can be equipped with the skills and mindset to identify, scale, and promote sustainable practices. Such an approach can transform education into a powerful conduit for unity and action in the face of environmental adversities.

Complex global challenges like climate change or biodiversity loss necessitate interdisciplinary approaches. By harnessing the knowledge from various fields—be it ecological sciences, socioeconomics, or political studies—solutions that are more holistic and resilient can be crafted. Interdisciplinary strategies can help mend the disconnects that arise from compartmentalized thinking, promoting a more seamless integration of sustainability into every facet of human life.

The exchange of traditional knowledge between cultures can be an invaluable asset in the sustainability toolkit. Indigenous and local knowledge systems have long understood the principles of living in harmony with nature. By respectfully integrating such knowledge with modern scientific approaches, humanity can benefit from centuries of wisdom and practices that are often overlooked in mainstream discourse.

Consumer choices and corporate practices form another arena where interconnectedness can influence global divides. By opting for sustainable products and services, consumers can send a powerful message to the marketplace, championing ethical practices and environmental stewardship. Similarly, corporations that recognize their role in society beyond mere profit can lead the charge in adopting responsible business models that reflect a commitment to planetary well-being.

At the heart of interconnectedness lies empathy and the realization that a shared destiny binds all life. When policymakers and leaders embody empathetic governance, reflecting their constituents' diverse needs and voices, more equitable and sustainable policies emerge. This form of governance, when it is sensitive to the symbiotic relationship between humans and nature, can be instrumental in bridging divides that otherwise inhibit collaborative progress.

Art and culture as expressions of human creativity and experience can also serve as compelling mediums to communicate and inspire action toward sustainability. Through storytelling, music, visual arts, and other creative expressions, the message of interconnectedness transcends spoken language, striking chords with universal themes and emotions that unite audiences around the globe.

Global challenges require global solidarity, and interconnectedness can be a catalyst for this unity. International accords and multilateral agreements serve as legal instruments and symbols of a collective commitment to a sustainable future. While these agreements are far from flawless, they are a testament to the power of cohesive action in the face of shared threats and a stepping stone toward more robust cooperation.

Finally, addressing global divides through interconnectedness aligns with the ethics of reciprocity—often found in many cultural and religious contexts as the "Golden Rule." This principle, which promotes treating others as one would wish to be treated, encapsulates the spirit of mutual respect and cooperation necessary to navigate and transcend the complex sustainability challenges of our time.

Unity through diversity, as espoused by systems thinking, dictates that our differences are not just strengths, but also opportunities for innovation and resilience. Diversity of thought, experience, and perspective can fuel the evolution of adaptive and flexible sustainability solutions tailored to fit the unique contexts of different communities while serving the collective interests of the global community.

Embracing interconnectedness, therefore, is not merely a strategy—it's a philosophy that underlines the imperative for a connected approach to sustainability. It is a reminder that in our capacity to work together lies our ultimate strength: the essence of human ingenuity and the key to overcoming the divides that scar our shared home. It is the acknowledgment that only through the sum of our collective actions can we aspire to a more sustainable and equitable planet for all its inhabitants.

In conclusion, the pursuit of sustainable living is inextricably linked to our ability to harness interconnectedness. The path ahead is strewn with challenges but also with hope, as every bond formed between individuals, communities, and nations contributes to a more sustainable and resilient world. It is this spirit of cooperation and shared purpose that will guide our steps as we strive to mend the global divides and ensure the baton of stewardship is passed on to future generations, galvanized and secure in the knowledge that they are part of a truly interconnected world.

Ethical Considerations in Global Interdependence

The fabric of our world is interwoven with threads of interdependence, each strand connecting distant cultures, economies, and environments. Now that we've explored the power of interconnectedness to foster sustainability and overcome divides, it's crucial to turn our focus to the profound ethical considerations that underpin this global tapestry. The decisions made in one corner of the world inevitably ripple across the seas, impacting lives and landscapes thousands of miles away. This interconnectedness comes with a moral responsibility to acknowledge the weight of our collective actions and choices.

In this dance of global interdependence, one must first recognize that our actions are not isolated; they have the potential to either harm or heal. Companies sourcing materials from far-off lands must consider their suppliers' working conditions and environmental stewardship. Focusing solely on profits is insufficient when human rights and ecological integrity hang in the balance.

Viewing climate change through the lens of ethics, we understand it not just as an environmental concern, but as a matter of justice. Those who contribute least to carbon emissions are often the ones most affected by its consequences, laying bare the inequities of shared resources. Thus, the onus falls on developed nations to lead in reducing impact and aiding vulnerable communities to adapt.

Moreover, ethical consideration demands respect for cultural heritage and knowledge. As we integrate technology and traditional practices, we must ensure the protection of the intellectual property rights of Indigenous communities, whose environmental wisdom is invaluable, yet often exploited without just compensation or acknowledgment.

The global marketplace offers boundless opportunities for cultural exchange, but also poses risks to cultural preservation. Artisans whose crafts have been passed down for generations find their designs copied and mass-produced, diluting the very essence that makes them unique. Ethical interdependence must champion fair trade and respect for the sanctity of cultural expression.

Investment and development, while powerful engines for economic growth, can't bulldoze through communities without considering their impact on local cultures and ecosystems. Responsible investing should prioritize projects that yield financial returns, support sustainable development, and empower communities.

To share technology and innovation ethically across borders is to democratize progress. But all too often, advanced technologies are inaccessible to those in less affluent nations. Addressing this divide means sharing innovation and collaborating to make it adaptable and affordable for varied contexts.

As consumers, our ethical footprint extends beyond the checkout counter. Mindful consumption practices necessitate awareness of a product's life cycle and the welfare of those who bring these goods to market. The demand for transparency and corporate social responsibility speaks to a growing consciousness among consumers who wish to make purchases that align with their values.

Regarding food, the path from farm to plate is intricate and globalized. Ethical interdependence in our food systems urges us to consider the sustainability of our diets, the plight of farmers and animals, and the preservation of biodiversity.

As we address the critical issue of water scarcity, ethical considerations in water management and sharing are vital. Water-rich nations must not commodify this precious resource at the expense of water-poor communities. Instead, collaboration and equitable policies should guide water governance.

In global health, pandemics demonstrate how quickly a health crisis can become a global crisis, emphasizing the ethical imperative to extend healthcare resources and knowledge across borders. Strengthening health systems everywhere is a prophylactic against the spread of disease, ensuring a robust defense for all.

Finally, the educational systems that nurture our youth must infuse global citizenship and ethical consciousness into their curriculums. Equipping future leaders with the knowledge and sensitivity to navigate an interdependent world ensures they will uphold the ethics of sustainability and respect diversity and equity.

The tapestry of our world demands we stitch together policies and practices that are economically viable and morally sound. Our sustainability efforts must be rooted in ethics that resonate with the heart as much as they make sense for the ledger. Only then can we craft a world that offers every inhabitant—no matter their geography—the dignity and opportunity they deserve.

Cultural Sustainability and the Web of Interconnectedness

The intricate web of interconnectedness within and across cultures plays a vital role in the quest for a sustainable future. It holds the threads that bind our collective knowledge, wisdom, and appreciation for the world. This section delves into the heart of cultural sustainability—a concept that sees the preservation and evolution of cultural diversity as crucial for a sustainable planet.

Cultural sustainability is not a standalone theme but an integral part of the broader sustainability conversation. It recognizes that cultural practices, beliefs, and values significantly impact the environment and our shared resources. As we embrace interconnectedness, the cultural sphere must be viewed as a fertile ground where sustainability can flourish.

For many cultures, the bond with nature is woven deeply into their identity. Such cultures often hold profound knowledge of local ecosystems that has been refined over generations, enabling them to live in harmony with their surroundings. In fact, Indigenous peoples have tenure rights over or manage a quarter of the world's land surface, which coincides with equally significant proportions of the world's biodiversity. By acknowledging and respecting these cultural relationships with the environment, humanity can glean valuable insights into sustainable living.

The web of cultural interconnectedness goes beyond environmental stewardship. It encompasses food systems, language, art, tradition, and other aspects of society. Each of these cultural elements has sustainability dimensions that, when thoughtfully considered, can contribute to the resilience and well-being of communities. For instance, local food practices often incorporate sustainable farming techniques passed down over time, reflecting a commitment to the land's and people's health.

Language, too, is a core component of culture with ties to sustainability. When a language becomes extinct, we lose more than words—we lose ecological knowledge and perspectives that are valuable for understanding and preserving our natural world. Therefore, sustaining linguistic diversity can be part of a holistic strategy to maintain biodiversity.

The expression of culture through art and music often carries powerful messages about the environment and our place within it. These creative avenues can inspire individuals and societies to reflect on their impact and their potential to enact positive change. By fusing sustainability messages with cultural art forms, ideas are shared in a way that resonates on a deeper emotional level.

Even in the realm of technology, there's an opportunity to forge links between culture and sustainability. Traditional knowledge and modern innovation can collaborate to create effective and culturally respectful solutions. This synergy can lead to technologies that enhance communities' sustainable practices.

It's not just about what we can learn from various cultures, but also about providing a platform for these diverse voices to be heard. Participation in global sustainability dialogue should be diverse and representative so varied perspectives can help shape policies that are both environmentally sound and culturally sensitive.

This interconnectedness also holds practical implications for approaching the exchange of knowledge and resources. Sustainable development initiatives become more impactful when they are co-created and consider cultural insights and respecting local contexts. Such collaborations can effectively bridge traditional knowledge systems with scientific approaches to address environmental challenges.

Addressing climate change, conserving biodiversity, and protecting the Earth cannot be achieved by a one-size-fits-all method. Each cultural setting offers unique approaches to sustainability that can inform global strategies. By weaving these diverse strands, a rich tapestry of sustainable practices emerges as a composite that is more resilient than any single thread could be alone.

The realization of interconnectedness drives an ethical imperative to recognize our shared responsibility. We must cultivate empathy and solidarity across cultural divides to work for the preservation of our planet. The perpetuation of cultural sustainability is not a choice but a necessity for the survival and enrichment of humanity's collective heritage.

As we learn from and support each other's cultural strengths, we open pathways for mutual growth and understanding. This shared journey toward sustainability transcends national boundaries, uniting us in purpose and action. It becomes not just a pursuit of environmental policy, but a movement grounded in the belief that our cultural diversity is as critical to our survival as biological diversity.

Finally, embracing interconnectedness in cultural sustainability means recognizing actions taken today echo into the future. By nurturing cultural continuities, we are planting seeds for future generations. It is an everlasting journey of learning, adaptation, and collective stewardship. Within this web of interconnectedness, every thread sustains another and, together, they sustain the whole.

In conclusion, cultural sustainability intertwined with the web of interconnectedness presents an essential framework for achieving a thriving, balanced, and resilient planet. Our cultural heritage, enriched by its diversity, is a repository of solutions and a source of inspiration for sustainable ways of living. Together, we can foster an environment where cultural and biological diversity are celebrated and protected, ensuring a legacy of sustainability for the world.

Chapter 5:
Global Issues, Local Colors

With the foundations laid in the preceding discussions on interconnectedness and diverse responses to environmental change, we venture into the vivid mosaic of localized approaches to global sustainability challenges in "Global Issues, Local Colors." Here, we examine the rich palette of measures communities around the globe take to address the urgent, universal crises of our age—most pressingly, climate change—while painting their solutions in the unique hues of their cultural and geographical contexts. Climate action is not a monochrome; it thrives on the diversity of its expressions, from the innovative agricultural practices of smallholder farmers preserving biodiversity and soil health to coastal cities adopting ancient stormwater management principles to combat rising seas.

This chapter illuminates how local knowledge systems, steeped in cultural heritage, are not relics of the past but living libraries that hold the keys to adaptation and resilience in the face of mounting global pressures. It's not merely enough to recognize the existence of such wisdom. We must actively integrate these practices into our modern strategies, focusing on how interconnected the fates of disparate regions truly are and ensuring the path to sustainability respects the colorful spectrum of human experience.

Diversity in Addressing Climate Change

In a world intricately woven with myriad cultural threads, addressing climate change necessitates harnessing this beautiful diversity to foster innovative and locally resonant solutions. Global communities perceive and interact with their environment through a unique cultural lens, which can be a powerful catalyst for creative strategies that honor traditional knowledge while embracing modern scientific breakthroughs. No one-size-fits-all remedy exists in this dynamic field. Instead, an array of multicolored, locally informed approaches paints a more sustainable future.

When coupled with cutting-edge technology, Indigenous practices can lead to a synergistic effect that amplifies resilience and sustainability at local and global scales. Encouraging a cross-pollination of ideas that respects the nuances of culture, this chapter posits that the way forward in our climate crisis is through a rich mosaic of practices as diverse as the ecosystems and cultures they aim to protect.

Traditions as a Source of Innovation

In the deep well of human traditions, we find time-tested patterns that have sustained communities for generations. As the preceding chapters have demonstrated, cultural diversity enriches our approaches to sustainability by offering many perspectives. Let us now immerse ourselves in the rich interplay between tradition and innovation and explore the potential for our inherited customs to spawn contemporary sustainable solutions.

Historic practices, woven skillfully into the fabric of various societies, have been the invisible hand guiding communities through the challenges of their times. These traditions, ranging from agriculture to community organization, are not merely artifacts of a distant past but serve as a source of inspiration for present-day innovation. They carry valuable insights into living harmoniously with nature, adapting to environmental changes, and utilizing resources responsibly.

Consider the ancient water management systems like the qanats of Iran or the stepwells of India. Both were sophisticated engineering marvels that served as oases in arid environments. These systems exemplify the ingenuity born out of necessity and provide a framework for modern sustainable water management in drought-prone areas.

Similarly, the Polynesian wayfinders traversed vast expanses of the ocean guided by an intimate understanding of star paths, winds, and marine life, showcasing a tradition of navigation that minimizes the need for modern technology. Today's interest in renewable energy harkens back to these forms of ancient wisdom, emphasizing adaptation and alignment with the natural world to create low-impact solutions to large-scale problems.

At the heart of many traditional agricultural practices, such as crop rotation and intercropping, lies a deep awareness of the land's health. Unlike industrial agriculture, these sustainable farming techniques prioritize maintaining soil fertility and controlling pests without chemical inputs. Current agricultural innovation often revisits these principles to develop regenerative farming practices.

Community-driven governance systems, rooted in traditions of collective decision-making and resource sharing, also inform today's efforts to build resilient local economies. New models that emphasize sustainability and equity can emerge by leveraging the social capital inherent in traditional sharing economies.

Traditional medicine, which fosters healing and wellness through natural means, points us toward a vast repository of knowledge about plant-based remedies and holistic health. This time-honored wisdom has the potential to complement and even shape modern healthcare and pharmaceutical research, which can lead to more sustainable health systems.

Traditional arts, including storytelling, dance, and textiles, can be powerful forces for innovation. They provoke critical reflection and dialogue on sustainable living, enabling communities to visualize and create a shared sustainable future.

The road to sustainability is not a hasty sprint; it's a transformative journey necessitating both reverence for the teachings of our forebearers and the courage to adapt their wisdom to our contemporary needs. This synthesis calls for a delicate balance between maintaining the essence and intent of traditional practices and molding them to address our emerging global challenges.

As we move forward, it is crucial to acknowledge that drawing from traditions is not about mere replication, but rather about interpreting their core values in a way that resonates with the present. In doing so, we honor the legacy of our ancestors and take an active role in extending the lineage of sustainable innovation. Initiatives that integrate traditional knowledge with modern technology can lead to developing new products or services with smaller ecological footprints. For instance, integrating biophilic design in urban development is inspired by traditional concepts of harmony with nature and could redefine modern architecture.

Furthermore, tradition-inspired innovations may hold the key to individual and community empowerment through enabling people to reclaim their heritage and strengthen their identity. These innovations can instill a sense of pride and autonomy, fostering the community cohesion necessary for the pursuit of sustainable development.

In conclusion, embracing the interconnectedness of our past, present, and future compels us to view traditions not as static relics but as dynamic catalysts for sustainable innovation. Through this understanding and application of inherited knowledge, communities worldwide are finding new and resourceful ways to thrive. The future of sustainability may very well hinge on our ability to recognize the wisdom we have already been given and to blend it thoughtfully with the ingenuity of the modern age.

As we continue this exploration of the roles culture and tradition play in charting a sustainable future, let us bear in mind that innovation is as much about rediscovery as it is about new discovery. By looking backward with respect and forward with hope, the tapestry of traditions can lead us toward innovative practices that uphold the principles of sustainability for generations to come.

Social Resilience: Lessons from Around the World

As we delve into the rich array of approaches to sustainability around the globe, one aspect inevitably stands out: resilience. Broadly defined, resilience refers to the capacity of a system—be it a person, a community, or an entire society—to recover from setbacks, adapt to change, and continue to develop. It's the indefatigable spirit that drives societies to not just survive, but thrive, in the face of adversity.

Worldwide, unique cultural, economic, and environmental contexts have given rise to equally distinctive forms of social resilience. These examples offer profound lessons on how communities can enhance their durability in the face of global challenges such as climate change, economic shifts, and societal transformations.

One such example is the tight-knit community structure found in rural Japan. Known as *kyosei*, this social fabric emphasizes mutual support and cooperation. When disaster struck in the form of the 2011 Great East Japan Earthquake, this interdependent culture enabled communities to organize rapidly and pool resources and expertise to rebuild their shattered landscapes and lives.

Similarly, the concept of *Ubuntu* in Southern Africa, which encapsulates the belief in a universal bond of sharing that connects all humanity, has seen communities there become exemplars of resilience. This philosophy encourages a collective approach to problem-solving to ensure no one is left behind during crises.

In Latin America, community resilience is often characterized by *buen vivir*, or "good living," a principle rising from Indigenous beliefs about living in harmony with one's environment. It champions sustainable living and prioritizes the well-being of the community over individual gains—a concept that has allowed communities to manage natural resources sustainably while also ensuring social equity.

Nordic countries, on the other hand, exemplify resilience through their emphasis on social safety nets and inclusive decision-making processes. These nations invest heavily in education, healthcare, and social services to create robust systems that act as buffers against economic and social shocks.

Moving to the Southeast Asian region, we find the *bayanihan* spirit of the Philippines, which manifests in communal work and cooperation, particularly in times of need. This tradition of mutual aid has seen communities unite to move entire houses to safer grounds during typhoons, showcasing incredible unity and resilience.

In the small island states of the Pacific, resilience has been crafted by necessity in the face of rising sea levels. To adapt, these communities have utilized traditional knowledge such as building homes on stilts and maintaining mangrove forests as natural protection against storm surges and flooding.

One of the undervalued aspects of resilience is social innovation, the creative processes through which societies can reconfigure their structures in response to external demands. For instance, Ghana's informal sectors, which serve as the backbone of its economy, display resilience through entrepreneurial spirit and adaptability, which allows communities to make the best of sparse resources.

Scandinavian design, known for its simplicity and functionality, also provides insight into the role innovation plays in resilience. Such design principles have resulted in community spaces that are not only aesthetically pleasing, but also adaptable to varying social needs and environmental conditions.

Resilience can also arise from adversity, as seen in post-conflict regions. The rebuilding process in communities within Rwanda has integrated the concept of *ubuntu*, which fosters reconciliation and a collective effort toward sustainable economic development and social cohesion.

In contrast to top-down initiatives, community-driven resilience has proved far more effective. Grassroots movements, often born out of necessity, address local needs with tailored solutions. From community energy projects in German villages to water conservation initiatives in Rajasthan, India, these self-organized efforts have been at the forefront of sustainable and resilient practices.

The stories of resilience that pepper the globe are as diverse as they are instructive. What's apparent is that the common thread weaving through each tale is a collective spirit—a shared understanding that survival is a group endeavor, and the focus on sustainable practices is ever more crucial in our interconnected world.

As businesses and policymakers strive to facilitate resilience at various scales, there's much to be learned from these examples. Central to their success are the recognition of the intrinsic link between human systems and natural ecosystems and the realization that social resilience is as much about fostering strong relationships as it is about economic stability or environmental stewardship.

As the final chapters of this exploration are written by none other than the actions we take today, it becomes clear that resilience isn't just about bouncing back; it's about bouncing forward with a vision for sustainability that is invariably colored by the local hues of culture, tradition, and innovation. It's through this lens that the lessons of social resilience gleaned from around the world become beacons of hope for a sustainable and vibrant future for all.

Chapter 6:
Sustainability and the Language of Culture

As we transition from the vibrant tapestry of local colors and global issues, this chapter delves into the intricate alignment between sustainability and the language of culture. At their cores, cultural narratives embody the transcendental power to turn ideals into tangible outcomes, bridging the ethereal gap between abstract values and concrete environmental actions. Through the vernacular of shared traditions, heritage, and collective memory, sustainability finds not only its voice, but also its operational blueprint. It's in this discourse that communities around the world interpret the very essence of sustainability: informing and reshaping behaviors in harmony with the Earth's cycles. This mutual enchantment between the environment and culture underscores the necessity of a language that echoes the global urgency while respecting the nuances of local dialects.

Through this chapter, a confluence of symbolism, tradition, and modern imperatives, we uncover how cultural lexicons guide us toward a future where sustainability isn't just preached, but practiced with a fervency that's as infectious as it is imperative for our shared survival. This isn't simply about translating words; it's about transmuting values into a global chorus that sings in the key of sustainable living. As we explore the multifaceted dialogues and stories that propel this movement, let us pause and listen to the ancestral whispers and contemporary echoes that guide us toward inching the world closer to enduring equilibrium.

Translating Values into Action

Previous chapters have laid a rich mosaic of cultural insights and conceptual paradigms. Now, we arrive at the precipice of change—the point at which values are beckoned to manifest in action. Within the cultural context of sustainability, one must ponder how deeply held principles and mores can morph into palpable initiatives and concrete habits. It's one thing to proclaim adherence to sustainability, another entirely to live it out loud.

Translating values into action is less about the grand gesture and more about the consistency of small, mindful movements. Each step informed by cultural understanding can cumulate in a stride toward enduring change. Actions must be both locally resonant and globally aware, walking the tightrope between specific cultural realities and the universal challenges mankind faces.

The power of action lies within its capacity to speak across linguistic and cultural barriers. When individuals opt for modes of transport that reduce carbon footprints or when communities choose to protect natural resources, they send a powerful message transcending words. In this global dialogue, sustainability becomes the common language.

Consider the traditional practices that have sustained societies for centuries such as reciprocity with the land, thriftiness in resource use, and reverence for nature, all of which are ripe for translation into modern sustainability efforts. By marrying these ancient tenets with contemporary needs, cultures around the world can fashion a future that's both fertile and feasible.

Community engagement is a force multiplier in this translation of values. It bridges the gap between individual conviction and collective action, tapping into the strength that comes from shared purpose. Collaborative projects like community gardens or local clean-up initiatives hinge on the principle that a tapestry of small acts can achieve large-scale transformation.

Education plays a pivotal role in this process. As we instill sustainable values within the next generation, we must also equip them with the tools to turn those values into everyday practices. Through a curriculum that embraces ecological literacy and hands-on learning, young minds are encouraged not only to understand the principles of sustainability but to live them.

Businesses and corporations aren't exempt from the call to action. On the contrary, the private sector has the resources and influence to make substantial impacts. Companies that embody corporate social responsibility and environmentally friendly practices not only set an example, but also can pivot the market toward a greener economy.

Policy also plays a critical role in actualizing cultural values. Government bodies that create and enforce environmental regulations are translating an abstract valuing of nature into a tangible safeguarding of it. In this context, policy becomes the canvas upon which societal values are painted, and a space is created where ecological respect is given regulatory teeth.

However, action is not immune to challenges. Each cultural context comes with its own set of complexities and constraints. For some, the obstacle may be a lack of resources; for others, resistance to change. Patience, persistence, and sensitivity to local circumstances are vital in overcoming these hurdles.

Effective translation of values into action also involves continuous reflection and adaptation. Monitoring and evaluation allow for the honing of strategies in real-time, ensuring actions remain aligned with intentions and outcomes reflect goals.

Technology, often seen as the bastion of progress, can either hinder or help. Its marriage with tradition must be a delicate dance where technological advancements bolster—rather than bulldoze—cultural practices geared toward sustainability. Digital tools can disseminate knowledge and foster collaboration, but they must be wielded with caution and conscience.

Moreover, the arts and storytelling serve as potent vehicles for enacting values. They inspire, provoke, and galvanize communities into action with a reach that extends beyond mere data and discourse by tapping into the emotive core of human experience.

Individuals, communities, and societies forge a shared destiny in the crucible of action. The daily choices we make—what we buy, how we travel, how we dispose of our waste—are the legible footprints of our values. They give measure to our commitment to a sustainable culture that celebrates the preservation of the Earth.

Thus, as we embark on this journey toward sustainability, we must not only carry our values in our minds, but also engrave them in our actions. It's a weaving of ritual, habit, and long-term thinking—a symphony of conscious choices that resonate with our deepest cultural truths and ripple out to shape the world's destiny.

As values translate into action, so too does hope transform into reality. Every compost pile turned, every tree planted, and every policy enacted is a testament to the potential embedded within cultural sustainability. It marks the transition from aspiration to actuality, from knowing better to doing better, from caring to conserving.

The Power of Narrative in Shaping Behavior

As our collective understanding of sustainability evolves, it is becoming clear that the language of culture has a profound influence on shaping human behaviors and attitudes. Within this intricate tapestry, the power of narrative stands out as an essential thread, a conduit through which cultural values, norms, and priorities are transmitted and internalized. Narratives are potent instruments; through stories, we define our identity, convey knowledge, share experiences, and fashion our perspective of the world.

In conveying the urgency of sustainability, narratives have the ability to transform abstract environmental concepts into relatable and actionable ideas. The stories we tell can frame sustainability in a way that resonates with our intrinsic motivations, thereby playing a crucial role in driving behavioral change. As such, narratives that encapsulate the balance of nature and human innovation are instrumental in promoting sustainable choices.

The cultural narratives that have been passed down through generations offer a rich repository of wisdom, which emphasizes connections between people and the planet. These tales often encapsulate a respect for natural resources and cycles, a reverence for conservation, and a spirit of cooperation. By sharing these stories, we reaffirm the principles that underpin sustainable living and rekindle a collective sense of stewardship over the environment.

Furthermore, contemporary narratives—spanning from literature and film to digital media—are equally critical in shaping society's behavior toward sustainability. Through engaging storytelling, these modern narratives highlight the consequences of unsustainable practices and inspire positive action. They play an educational role through broadening the audience's understanding of complex environmental issues by humanizing them, making them relatable, and crafting a call to action that aligns with their cultural values.

When narratives incorporate elements of empathy and ethical consideration, they become particularly effective in altering behavior. By eliciting emotional responses, stories can transcend intellectual barriers and foster a heartfelt commitment to sustainable practices. Through empathy, individuals can truly appreciate the impact of their actions on others and on future generations.

The role of narrative is not limited to the transmission of ideas; it also provides a platform for community engagement and social cohesion. Sustainability crises are not experienced in isolation—they are collective challenges that require collaborative solutions. Shared narratives can unite disparate groups by reinforcing common goals and highlighting the interconnected nature of social, economic, and environmental well-being.

Moreover, drawing on cultural archetypes and symbols within a narrative can reinforce the importance of sustainability. Such archetypal imagery, deeply ingrained in the collective psyche, can swiftly convey complex ideas and encourage a rethinking of individual and collective roles in the quest for a more sustainable future.

There is also a need for narratives that empower and embolden those who may feel disconnected or powerless in the face of global sustainability challenges. These stories inspire others to believe in their ability to effect change and encourage active participation in sustainability efforts by elevating the voices of community leaders, activists, and everyday heroes.

Critical to the crafting of effective narratives is ensuring they are culturally sensitive and inclusive. Recognizing the diversity of experiences and values across different cultures is vital for creating stories that are accessible and engaging for all audiences. This inclusivity not only enriches the narrative, but also paves the way for a more pluralistic and robust sustainability movement.

Furthermore, the narrative approach to promoting sustainability offers the advantage of adaptability. As conditions change and new challenges arise, stories can evolve to address these shifts by providing fresh perspectives and innovative solutions. This flexibility is essential in maintaining momentum and continued relevance in sustainability discourse.

To truly harness the power of narrative in sustainability, there is also a need for strategic storytelling. This involves identifying the most effective channels for dissemination, whether through traditional media, social media platforms, educational settings, or other forms of communication. The strategic use of narratives can maximize their reach and impact, fostering broad-based engagement and support for sustainability initiatives.

Moreover, the successful application of narrative in shaping behavior toward sustainability hinges on authenticity. Authentic stories that are based on real experiences and evidence resonate more deeply with audiences and have greater credibility. They not only inform, but also invite listeners to trust and invest in the message.

The power of narrative is also evident in its ability to challenge existing norms and inspire a vision of what could be. Sustainability necessitates a reimagining of societal paradigms, and narratives that paint a picture of a desirable and attainable sustainable future can motivate individuals and communities to strive toward that vision with renewed purpose.

As guardians of these narratives, storytellers have a profound responsibility. They are the craftspeople who shape the lens through which we view our relationship with the world and our potential to foster a more sustainable existence. By weaving narratives that are as compelling as they are instructive, they can steer cultural consciousness toward a future where sustainability is not just an aspiration but a lived reality.

Having dissected the power of narrative in shaping behavior, it's evident that the tapestry of sustainability is incomplete without the vibrant threads of stories that span the breadth of human experience. Narratives are more than mere words; they are the catalysts for change and the means through which we understand, participate in, and ultimately transform our world. In the journey toward sustainability, the stories we choose to tell—and live—will chart the course of our collective destiny.

Chapter 7:
Indigenous Wisdom for Modern Challenges

A s we turn the page to "Indigenous Wisdom for Modern Challenges," we delve deep into the timeless repository of Indigenous knowledge to examine how these ancient streams of understanding can offer powerful solutions to contemporary ecological dilemmas. For millennia, Indigenous cultures have honed practices finely attuned to their environments, fostering resilience, sustainability, and a profound sense of stewardship for the earth—a stark contrast to modern society's quick-fix, transactional mentality.

This chapter illuminates the wisdom held by Indigenous peoples, from land management techniques that enhance biodiversity to philosophies that promote living with a light ecological footprint. In an age when climate change, biodiversity loss, and resource depletion threaten our collective future, we explore how Indigenous practices not only conserve ecosystems, but also anchor cultural identities. It's a compelling call to integrate this wisdom into modern environmental management and policymaking frameworks in order to create a symbiotic alliance for a sustainable future. As we untangle the complex web of global challenges, embracing Indigenous knowledge isn't just beneficial—it's imperative.

Timeless Knowledge in a Time of Crisis

As our world grapples with unprecedented challenges, from the throes of climate change to the convulsions of global health crises, we need the guidance of diverse knowledge systems more than ever. One rich reservoir of wisdom lies amidst the practices and convictions of

Indigenous peoples whose intimate relationship with the natural world offers unparalleled insights for survival and resilience. Across diverse landscapes, these communities have, for millennia, honed their ability to live sustainably, adapting ingeniously to shifting environments. Yet, it is precisely in times of crisis that the relevance and value of Indigenous wisdom can be felt most powerfully and integrated into the fabric of global solutions.

The wisdom of Indigenous cultures is steeped in a profound understanding of ecological systems and the delicate balance of life. Often passed down through generations in stories and traditions, these teachings encode survival techniques, philosophy, ethics, and a way of being that prioritizes harmony with nature. Their relevance becomes stark in the face of crises that, more often than not, are exacerbated by modern society's disconnect from the natural world. This intrinsic connection to the Earth imbues Indigenous practices with timeless relevance, especially as modern society seeks to reconcile progress with planetary boundaries.

A critical aspect of Indigenous wisdom is the concept of interconnectedness. It dictates that every action has a ripple effect on the environment, and this holistic understanding can guide us in creating systems that are both sustainable and equitable. For example, traditional knowledge systems can inform sustainable agriculture practices that not only improve food security, but also enhance biodiversity and soil health, such as the intercropping methods practiced by many Indigenous agricultural communities.

Furthermore, the adaptive capacity of Indigenous knowledge becomes most apparent during a crisis. These communities are often on the frontlines of environmental change, prompting them to develop responsive strategies that ensure their survival. As such, they become invaluable case studies and partners in crafting resilient responses to global challenges. By listening to and respecting Indigenous voices, policymakers and researchers can gain insights into managing ecosystems under stress and developing adaptive climate strategies grounded in millennia of experience.

However, utilizing Indigenous knowledge in the modern context is not without its challenges. There is a significant duty to approach these traditions with respect, ensuring their survival and integrity while resisting exploitative or extractive tendencies. It requires a partnership in which Indigenous peoples are not merely subjects of study but active contributors, decision-makers, and beneficiaries in the processes that govern the use of their knowledge and lands. It is crucial that their intellectual property rights and cultural significance are rigorously protected and promoted.

It's imperative to recognize the systems of governance inherent in Indigenous communities as models for sustainable and participatory management. The collective decision-making processes often employed by these groups exemplify inclusive governance that allows for a broad range of voices to be heard—a practice modern democracies can learn from, especially when addressing shared crises.

Moreover, the resilience of Indigenous knowledge in crisis situations often hinges on the strength of social networks and cultural solidarity. These social constructs foster community resilience, enabling groups to endure and adapt to hardships collectively. Modern society's tendency toward individualism often undermines such solidarity, but adopting models of collective action derived from Indigenous practices can strengthen community bonds and resilience.

Crises can also serve as an opportunity to rectify historical injustices by reasserting the importance of Indigenous rights and steering away from the narratives that have marginalized Indigenous peoples and their knowledge. Such efforts should involve restoring access to traditional lands and resources, which are critical for continuing cultural practices and sustaining Indigenous communities.

One cannot ignore the spiritual dimensions of Indigenous wisdom, which often guide their resource management and conservation ethics.

These spiritual beliefs underscore the innate value of nature beyond utilitarian considerations—something that modern conservation efforts sometimes overlook, much to their detriment. Embracing this spiritual perspective can deepen our sense of responsibility toward the Earth and foster conservation strategies that are as much about reverence for life as they are about scientific management.

The practice of reciprocity, another tenet of Indigenous knowledge, teaches us to give back to the Earth as much as we take. This reiterates the need for sustainable consumption and production patterns that modern economies must adopt. Living within our means is not just an Indigenous philosophy but a necessity for ensuring the viability of our planet for future generations.

The stories and experiences of Indigenous peoples also present valuable lessons in the art of resilience. They demonstrate how to maintain cultural identity and community coherence in the face of external pressures. In an era when globalization threatens to homogenize and erase unique cultural expressions, maintaining diversity is both an act of cultural preservation and resilience building.

As these traditional knowledge holders are increasingly threatened by environmental degradation, forced displacement, and assimilation, there is a moral imperative to safeguard and revitalize this wisdom. The international community's role in recognizing and upholding the rights of Indigenous peoples cannot be overstated. Their participation in global dialogues surrounding sustainability is not only fair but essential. Including Indigenous perspectives can enrich our collective understanding and enrich the global tapestry of knowledge.

In the synthesis of ancient wisdom with contemporary science, we find a potent strategy for addressing the multifaceted crises of our time. This entails creating transdisciplinary platforms where Indigenous knowledge is not marginalized but celebrated and integrated into the scientific community and policymaking processes.

The path forward, then, is not a rejection of modernity, but rather an integration of the old and the new. It's about listening to the Earth with as much intent as we listen to the latest scientific findings and viewing sustainability through the time-honored practices of those who have thrived in balanced coexistence with their environment. As we navigate these tumultuous times, embracing the timeless knowledge of Indigenous peoples can guide humanity toward a future where respect for the Earth and all its inhabitants is woven into the very fabric of our societies.

Timeless Indigenous wisdom, when adopted thoughtfully, holds the promise of a more resilient and compassionate world—one where crisis need not be a harbinger of despair, but rather a catalyst for renewing our bonds with nature and one another.

Addressing Threats to Indigenous Knowledge

In our pursuit of modernity, it's all too easy to overlook the wisdom Indigenous knowledge systems hold such as the deep understanding of ecology, community, and sustainable living honed over millennia. Yet, as the previous chapters have illuminated, the contemporary world faces challenges that this intrinsic wisdom can help address. However, before we can integrate this wisdom into broader strategies, we must first address and mitigate threats to the survival of Indigenous knowledge itself. This is imperative not just for the sake of cultural preservation, but for the very sustainability of our planet.

One of the most significant threats to Indigenous knowledge is the loss of languages. As languages become extinct, the nuances and specific terms related to traditional ecological knowledge and practices are also lost. This impoverishes our global cultural heritage and eliminates the potential for these systems to inform and improve contemporary environmental management strategies. Efforts to preserve Indigenous languages are thus critical, not only as a matter of respect for those cultures, but as a repository of ecological wisdom.

Another threat is the marginalization and displacement of Indigenous communities. As these populations are forced to relocate due to environmental degradation, urban expansion, or conflict, the connection between the people and their ancestral lands is severed. This break disrupts the transmission of knowledge between generations and often leads to the dilution or loss of cultural practices that can contribute to sustainability.

Intellectual property rights also present a contentious issue. The commodification of Indigenous knowledge and biopiracy—where companies patent Indigenous plants and knowledge without proper compensation or acknowledgment—threatens the control that Indigenous communities hold over their traditional knowledge. Developing legal frameworks that protect the proprietary rights of Indigenous communities can provide some measure of defense against such exploitation.

Educational systems that do not recognize or value Indigenous knowledge further perpetuate its erasure. Mainstream education often prioritizes Western epistemologies and methodologies, which leads to an educational gap wherein traditional knowledge is neither taught nor respected. This undermines the roles of Indigenous elders and educators as custodians and transmitters of knowledge, further endangering the continuity of traditional knowledge.

Environmental change and biodiversity loss have catastrophic effects on Indigenous knowledge, as many traditional practices are intimately connected to specific species and ecosystems. Climate change, habitat destruction, and pollution can all disrupt the delicate balance Indigenous practices have struck with their environment, thus undermining the viability of these practices.

Resisting these threats requires proactive measures. Cultural safeguarding involves the documentation of Indigenous languages, traditions, and environmental practices. It requires investing in community-led projects that strengthen the transmission of knowledge, such as cultural centers, educational programs, and sustainable community enterprises that showcase Indigenous methods of agriculture, resource management, and craftsmanship.

Land rights are central to the protection of Indigenous knowledge as well. Ensuring Indigenous communities have legal title to their ancestral lands prevents displacement and allows for the continued practice of traditional land management techniques that have sustained biodiversity for generations.

Institutionalizing Indigenous knowledge within international legal frameworks offers a pathway to substantial protection. Policy initiatives that recognize the value of Indigenous wisdom in biodiversity conservation and climate change mitigation can empower those communities. This involves respecting Indigenous intellectual property through mechanisms that ensure benefit-sharing and fostering their involvement in decision-making processes.

While often seen as an agent of modernization at odds with traditional practices, technology can be a powerful ally in preserving Indigenous knowledge. Digital archives, databases, and multimedia resources can immortalize the knowledge of elders and make it accessible for future generations. Moreover, new technologies can assist in monitoring environmental changes, protecting biodiversity, and disseminating traditional knowledge globally.

Aligning Indigenous knowledge systems with scientific research can provide reciprocal benefits. Collaborative approaches can respect and integrate traditional understanding with scientific methods, leading to more holistic environmental management strategies that draw on the best of both worlds.

Addressing threats to Indigenous knowledge is not a task that falls only to the communities that hold the knowledge; it is a collective responsibility. Consumers and businesses are called to promote products and services that respect Indigenous rights and contribute to preserving their cultures. Ethical consumption also plays a role in reducing harmful environmental practices that threaten Indigenous ways of life.

The engagement of international bodies and NGOs can help support Indigenous claims and foster global awareness about the importance of Indigenous wisdom. Publishers, educators, and media outlets can amplify Indigenous voices in order to ensure these perspectives are heard globally and factored into global conversations on sustainability and culture preservation.

Lastly, expressions of solidarity, such as purchasing from Indigenous producers, participating in cultural exchanges, and supporting Indigenous-led conservation initiatives, can have a profound impact. Interconnectedness goes beyond shared challenges; it encompasses shared growth and mutual learning. By standing with Indigenous communities, we demonstrate a commitment to their survival and a sustainable and resilient world for all.

As we contemplate the roadmap toward a sustainable future, the threats to Indigenous knowledge are inseparable from broader environmental and social challenges. Countering these threats requires a multidimensional approach that engages the culture, economy, policy, and technology. It demands sensitivity, respect, and the willingness to learn from those who have honed a symbiotic relationship with the Earth. Only by protecting and integrating these knowledge systems can we hope to weave a global tapestry robust enough to meet the modern challenges we face.

Integrating Indigenous Practices

As our journey through Indigenous wisdom and its relevance to modern challenges continues, we delve into the heart of cultural convergence, where ancient customs meet contemporary needs. In this exploration, we understand just how crucial integrating Indigenous practices is to achieving a harmonious and sustainable future. But how exactly can we weave these time-honored traditions into our present-day fabric without distorting their essence?

Firstly, it is essential to recognize Indigenous practices are not relics of a bygone era, but are living systems that have evolved with the ebb and flow of natural cycles. They exhibit a profound understanding of local ecologies. For instance, consider Indigenous agricultural methods that depend on intimate knowledge of the land rather than on synthetic inputs. These techniques, which often include crop rotations, polycultures, and agroforestry, can be adopted to bolster sustainable agriculture and enhance food security in our changing climate.

Secondly, Indigenous knowledge systems often prioritize long-term ecological balance over short-term gains. This contrasts starkly with the sometimes myopic focus of modern industry. By integrating such a perspective into business models and development policies, we can shift toward practices that are financially, environmentally, and socially sustainable.

Moreover, Indigenous practices in resource management, such as the controlled use of fire to maintain ecosystem health or the communal decision-making processes governing resource use, can inform contemporary natural resource management strategies. Such integration supports biodiversity and can lead to more resilient landscapes.

Integration, partnerships, and platforms for dialogue between Indigenous peoples and policymakers at all levels must be fostered in order to actualize these ideals. This requires both humility, respect, and the recognition of Indigenous rights and sovereignty while adopting an inclusive approach that genuinely values Indigenous expertise.

Additionally, it is crucial to defend and promote the use of Indigenous languages, as they carry nuanced understandings of the environment not easily translated into dominant languages. Protecting these languages goes hand-in-hand with conserving biodiversity, for their insights can guide the stewardship of natural resources.

Education also plays a pivotal role. Schools, universities, and informal learning platforms can incorporate Indigenous perspectives into their curricula. This can provide a more comprehensive understanding of ecological processes and foster greater cultural appreciation.

An ethical consideration is the avoidance of appropriation and exploitation of Indigenous knowledge. Partnerships should be built on consent, equitable benefit-sharing, and intellectual property rights that protect Indigenous cultural expressions.

Furthermore, integrating Indigenous practices into modern medicine holds promise for public health advancements. Many conventional pharmaceuticals have origins in traditional remedies, and a collaborative approach to ethnomedicine can unlock new healing possibilities while providing livelihoods and preserving knowledge.

We must not forget the urban context where the majority of the world's population now resides. Urban planning and architecture can draw inspiration from Indigenous concepts of space, community, and the environment to create cities that are not only efficient, but also nourish the human spirit.

Beyond traditional practices, Indigenous philosophical views, such as those that see humans as part of a larger community of life, can radically transform our worldview by encouraging greater respect for our shared environment and promoting ethical stewardship.

Lastly, conservation efforts benefit greatly from Indigenous participation. Parks and protected areas managed with, or entirely by, Indigenous communities frequently have better conservation outcomes. By acknowledging and supporting the role of Indigenous guardians, we can advance global conservation goals.

Embracing the integration of Indigenous practices is not merely an act of revivalism; it is an investment in viability. It is a recognition that these wisdoms, when woven into the fabric of modernity, render us all richer in our collective pursuit of a sustainable future.

Chapter 8:
The Role of Religion in Green Living

n our search for sustainability, we acknowledge that the spiritual dimension of human experience plays a crucial role in shaping our relationship with the Earth. The values, narratives, and practices enshrined within various religious traditions offer profound insights into the ethical treatment of the natural world.

This chapter delves into religion's potent influence on green living, where faith-based initiatives drive environmental stewardship, and moral imperatives issue a clarion call to action for believers and faith communities alike. Key scriptural teachings across different faiths often echo the importance of protecting Creation, and these teachings can catalyze action that respects the integrity of our planet.

The marriage of religious conviction with environmental ethics offers an empowering framework for action and encourages communities to embody sustainable practices that resonate with their spiritual beliefs—a strategy that not only appeals to one's sense of duty, but also taps into the collective power of shared values to foster a healthier, more sustainable world. Hence, religion can bridge the gap between understanding the need for environmental protection and actualizing it through everyday practices, thus guiding its adherents toward a more harmonious and meaningful engagement with nature.

Faith-Based Initiatives for Earth Stewardship

Across the rich mosaic of faiths that grace our planet, the call to stewardship of the Earth is a singular thread that weaves throughout. In this chapter, we will examine a variety of faith-based initiatives aimed at nurturing and protecting our natural world through the lens of diverse religious practices. It's not just about belief; it's about action.

In the Christian tradition, the concept of stewardship is rooted in the Book of Genesis, where humankind is granted dominion over the Earth (Genesis 1:26). This dominion is often interpreted as a sacred responsibility to care for creation, rather than a license to exploit it. Initiatives such as the Evangelical Environmental Network seek to mobilize Christians to tackle climate change and protect the environment, and advocate for policies that reflect their caretaker role.

Similarly, Islamic teachings emphasize the idea of humans as Khalifah—caretakers or vicegerents of Allah on Earth (Qur'an 2:30). EcoIslam is a movement that encourages Muslims to live sustainably and in accordance with the principles of conservation inherent in the Qur'an and the Hadith. They propagate the belief that environmental preservation is central to the faith and promote practices like water conservation, ethical consumerism, and renewable energy adoption within Muslim communities.

Judaism is no stranger to environmentalism either. Concepts such as Bal Tashchit—do not waste or destroy—offer a scriptural basis for contemporary Jewish environmentalism (Deuteronomy 20:19–20). This has given rise to organizations such as the Coalition on the Environment and Jewish Life (COEJL), which engages Jewish communities in efforts to confront climate change through a faith-informed lens.

The Hindu tradition venerates nature through the worship of natural elements and the understanding that the Divine permeates all Creation. The Bhumi Project, inspired by Hindu values, launched initiatives to reduce carbon footprints and enhance education regarding climate change within the global Hindu community.

Buddhism advocates for a deep respect for all living beings and the concept of interconnectedness with the environment. The notion of interdependence is exemplified in the Buddhist climate change statement to world leaders, which called on global leaders to acknowledge the environmental crisis and act with wisdom and compassion.

Among Indigenous faiths, a spiritual connection to the land and an intricate knowledge of ecological systems have always been intrinsic to their way of life. Initiatives such as the Indigenous Environmental Network (IEN) amplify these voices, promoting sustainable practices grounded in Indigenous spirituality and knowledge systems.

While different in rites and practices, Sikhism has theoretical roots in the stewardship of the Earth, too. Echoing this, the EcoSikh organization emerged as part of a long-term plan to address environmental issues from a Sikh perspective through engaging communities in tree planting and sustainability education initiatives.

These initiatives and organizations often partner with secular environmental groups to multiply their impact. Take, for instance, the Faith for Earth initiative by the United Nations Environment Programme (UNEP), which emphasizes the role of faith organizations in contributing to environmental consciousness and action.

Furthermore, interfaith coalitions like the Interfaith Rainforest Initiative are vital for uniting people from multiple faiths to protect the world's rainforests and the rights of Indigenous peoples who serve as their guardians.

Climate change, recognized as a moral issue, brings faith groups into direct dialogue with policy. Faith leaders are increasingly present at global forums such as the United Nations Climate Change Conferences, where they advocate for ethical approaches to environmental policy and remind world leaders of the moral imperative to act.

Education plays a significant role in these faith-based efforts. Many religious organizations have created curricula that integrate sustainability principles with theological teachings to inform and inspire future generations to care for the environment.

Pilgrimage, a practice shared by many religions, is also being reimagined as a force for good for the planet. The Green Pilgrimage Network encourages sustainable practices among pilgrims and the cities they visit in order to reduce environmental impacts while deepening spiritual experiences.

Lest we forget, the religious sector is a large landholder and is responsible for the management of extensive tracts of forest, agricultural land, and real estate. Faith-based groups leverage this to lead by example by implementing green building standards, renewable energy projects, and sustainable management of lands under their care.

To conclude, faith-based initiatives for Earth stewardship are as diverse as the religions that inspire them, yet they share a common goal: to foster a harmonious relationship between humans and the natural world. Religions, with their deep roots in communities and their capacity to inspire change, play a pivotal role in promoting sustainable living and ecological justice.

Moral Imperatives: A Call to Action

Within the tapestry of faiths and beliefs that have cradled our civilizations, there lies a profound source of moral imperatives that can ignite and fuel our pursuit of green living. Religious doctrines around the globe offer a singular vantage point, a call to a higher responsibility that transcends creed and culture. It's within this realm we realize the essentiality of integrating spiritual motivation with environmental action. This unity presents a clarion call for proactive stewardship of our shared planet, encapsulating not just the physical actions we must undertake, but also the ethical underpinnings that compel us toward them.

The cries of our beleaguered Earth are reaching a crescendo that can no longer be ignored. Religion, a beacon of moral guidance for millions, distinctly positions itself as a catalyst for ecological change. Holy texts and spiritual teachings have often inscribed in their core messages the virtue of caretaking the Earth and all its inhabitants. It's incumbent upon us to heed these transcripts not as archaic literature, but as contemporary blueprints directing us toward sustainability.

Green living is not merely a practical maneuver in the game of survival, but a moral expedition charted by the very tenets of faith. The sentiment of interconnectedness, professed by spiritual philosophies, ushers in an ethical imperative to respect and protect the ecological web of life. A genuine comprehension of this connectedness demands an active response and a shaping of lifestyles that honors those sacred bonds.

Each tradition calls its adherents to reflect their reverence for creation by embodying principles of conservation and restoration in daily life. To consider oneself a person of faith one must, by necessity, encompass the conscious decision to live in a manner that minimizes harm to the environment. This convergence of spirituality and ecological responsibility creates a powerful synergy for environmental advocacy and action.

However, faith-inspired green living isn't merely about personal practices; it's about galvanizing communities to collective action. Religious institutions hold immense influence and reach within society. They can contribute significantly to promoting sustainable practices, disseminating environmental education, and fostering communal eco-projects. When local faith leaders become ambassadors for sustainability, they can transform entire communities and inspire grassroots revolutions for green living.

In manifesting green living, faith communities shouldn't only focus on conservation, but also on restoration. It isn't enough to halt further damage; there must be efforts to restore what has been degraded. Ecosystem restoration, as a spiritual offering, honors the intrinsic value of nature's systems and recognizes humankind's role in healing the environment, rather than solely exploiting or dominating natural resources.

Our moral imperatives also demand a critical review of consumption patterns. Excessive consumerism has brought us to the brink of environmental catastrophe. Religions often promote a lifestyle of simplicity and moderation, which can inherently reduce ecological footprints. Embracing these values and making ethical consumption choices speaks directly to the moral soul of environmentalism, where every purchase reflects a principle, and every resource utilized holds a reverence.

The recognition of climate change as a moral issue has evolved within religious discourses, highlighting it as an environmental crisis and a question of justice. It is the vulnerable, often the least responsible for ecological degradation, who suffer its worst effects. Addressing climate change through the prism of ethics adds weight to the argument for urgent action and equitable solutions that align with religious precepts of compassion and fairness.

To actualize the moral imperatives of religion in the pursuit of green living, there is also a need to engage in interfaith dialogues and collaborative efforts. No faith exists in isolation, and the environmental challenges we face do not respect religious boundaries. Cooperative endeavors that cross faith lines can harness a broader base of support for sustainability initiatives by emphasizing our common ground: the Earth we all share.

The quest for sustainability must also transcend mere individual action and address institutional and systemic changes. Faith-based organizations can use their collective voices to advocate for environmental policies that align with moral values and influence the direction of local and international governance toward a more sustainable trajectory.

The moral imperatives of religion in green living hold the potential to reframe our cultural narrative. The narratives through which we understand our relationship with the planet can shift from dominance and detachment to stewardship and empathy. By centering our drive for sustainability around moral imperatives, we transform the ecological movement into one that is deeply human and inherently ethical.

In the spirit of moral imperatives, we're prodded to explore education as a crucial conduit for change. Faith-based communities have an opportunity to educate their congregations about the sacredness of nature and the practical ways to engage in conservation. This education should not be confined within the walls of houses of worship but should extend to the wider community in order to foster an ethic of care among all citizens, irrespective of their faith identities.

Lastly, the moral imperatives inherent in religious teachings prompt a reflective consideration of our legacy. What do we wish to bequeath to our children and future generations? Our actions today are drafting the heritage of tomorrow. We are accountable for creating a legacy reflecting respect, care, and thoughtfulness toward our planetary home—a living testament to our spiritual principles.

As we advance in this critical journey of assuming responsibility for our environment, let it be undergirded by the profound moral imperatives our various religious traditions espouse. This synthesis of faith and action carries the hope of a sustainable future imbued with respect, justice, and love for the Earth and all that dwells within it. These are more than mere calls to action; they are summons to live out the deep ecological wisdom embedded in our faiths with passion and purpose.

Chapter 9:
Art, Music, and the Rhythms of Sustainability

C ontinuing from the vital discussions on diversity and values, this chapter delves into the expressive forces sculpting society's eco-consciousness. Here, we uncover the profound impact of the arts as laboratories for innovation, where creativity intertwines with effective environmental stewardship to resonate change beyond words.

It's not just about aesthetics; art and music are instruments for transformation, amplifying the subdued voices of nature, inspiring communities to act, and engraving the essence of sustainability within the cultural lexicon. This symphony of artistic endeavor equips us with the emotive depth to shift perceptions, where each brush stroke and note in a melody becomes a heartbeat syncing with the Earth's own rhythms. In manifesting visions of balance and harmony, these mediums lend a soul to the discourse, vital for weaving the ethos of conservation into the fabric of daily life. As we explore their dynamic roles, we're not just spectators but participants in a dance of reinvention, fostering a resilient culture that can navigate the uncertainties of an evolving planet.

The Arts as Catalysts for Change

Throughout history, the arts have been a reflection of the times—a mirror to society's triumphs, tragedies, and transitions. As we continue to navigate a world increasingly shaped by the threats of environmental degradation and climate change, artists have become instrumental in catalyzing action. They wield the power to move minds, evoke emotional responses, and inspire change with a resonance that transcends boundaries, languages, and cultures.

The realm of art is vast and diverse, encompassing traditional visual arts, music, dance, literature, and emerging forms that blend various media. These artistic expressions have become pivotal in communicating sustainability's urgent narrative. Take, for instance, visual artists who depict the beauty and fragility of endangered ecosystems. Their work invokes a sense of urgency to protect these natural treasures.

Music, with its universal language, harmonizes with the rhythms of the natural world and the needs of society. It can unite people for environmental causes, as seen in benefit concerts for conservation or albums dedicated to raising awareness about climate change. Musicians like those participating in the environmental movements provide a soundtrack for change, which motivates communities to strive for a greener future.

Dance, another transformative art form, expresses complex ideas about nature and interconnectivity through the human body. Through performance, dancers can embody the ebb and flow of ocean tides or mimic the growth patterns of plants, visually interpreting the essence of sustainable living through their movements.

Literature, too, commands a role in fathoming sustainability's depths and dilemmas. Writers craft stories and narratives that can alter perceptions, challenge the status quo, and encourage readers to imagine sustainable futures. The dystopian and utopian novels forewarn the consequences of unsustainable practices and envision a harmonious coexistence with nature.

Street art and public installations have a unique ability to transform the civic spaces we inhabit into thought-provoking exhibitions. These works often interact directly with the environment, weaving the message of sustainability into the fabric of everyday life and making it accessible and unavoidable for a broader audience.

Theatre and performance art enable communities to explore sustainability by dramatizing environmental conflicts, societal challenges, and historical ecological relationships. Through the lens of performance, audiences can witness and reflect upon the impacts of their lifestyle choices and consider alternative, more sustainable behaviors.

Film and digital media offer powerful platforms for storytelling and have become key tools in shaping public understanding of environmental issues. Documentaries bring the plight of remote ecosystems directly to viewers' screens, while fictional films use narrative to engage viewers with environmental themes in relatable contexts.

Art can also serve as a bridge between scientific understanding and public perception. By making complex data visually compelling, artists can help demystify scientific findings, making the intricate relationships within ecosystems and the implications of climate change more comprehensible.

Cultural festivals centered around arts and sustainability provide opportunities for immersive experiences that celebrate eco-friendly practices and raise awareness of conservation efforts. These events can foster communities of like-minded individuals who share a passion for the arts and a commitment to sustainable living.

Education systems that integrate arts with environmental curricula can nurture a new generation that values creativity and is passionate about sustainability. Art projects focused on recycling, conservation, and biodiversity can instill habits of environmental stewardship in young minds, promoting a more sustainable future from early childhood.

In the realm of policy, art has the potential to influence lawmakers and stakeholders. Provocative public installations, compelling visual data narratives, and heartfelt performances can draw attention to environmental issues that are often sidestepped in political discourse, compelling change at the systemic level.

For communities grappling with the effects of climate change and environmental degradation, art offers a means of coping and rebuilding. By engaging with the arts, communities can process their experiences, share their stories of loss and recovery, and foster a shared vision for a sustainable path forward.

Arts also embody the practice of sustainability in their own life cycle. The trend toward eco-friendly art materials, sustainable event management, and conservation-minded artistic practice helps reduce the environmental impact of art itself, setting an example for other sectors to follow.

Finally, art can be an anchor for collective memories and identities, reminding us of what is at stake if we neglect the health of our planet. Artistic portrayals of the past—how communities have interacted with their environments or how landscapes have changed over time—can offer poignant reminders of the need to preserve the diversity and vibrancy of the world for future generations.

Indeed, the arts stand as powerful allies in the quest for a sustainable future. Through artistic endeavors, we find inspiration, engage with complex emotions, and grapple with the grand challenges our planet faces. Through creativity, we can envision a world where sustainability is not merely an aspiration but a rhythm that orchestrates the symphony of human existence.

Harmonizing Creativity with Conservation

The interplay of art, music, and sustainability creates a harmonious symphony, signifying a vision of conservation that speaks to the soul. In the byways of history, art and music have been crucial in expressing our relationship with nature, echoing a resonance of conservation that predates modern environmentalism. Today, these mediums hold the potential to transform sustainability from concept to experience, educating and influencing society's heartstrings toward a more sustainable tomorrow.

Yet, the nexus of creativity and conservation is more than merely theoretical; it manifests vividly when musicians use repurposed materials to craft instruments or when artists opt for sustainable materials. Each creation, be it an upcycled sculpture or an eco-conscious melody, acts as an ambassador for environmental values, embodying the essence of using less to create more.

The power of this fusion lies in its inclusivity and accessibility. Artistic endeavors do not require technical expertise in sustainability to partake in or appreciate. Instead, they invite individuals of all ages and backgrounds to engage with conservation issues on an intimate level and weave understanding and conviction into the fabric of everyday life. This communal approach to creativity engenders a collective commitment to stewarding our resources responsibly.

How we infuse sustainability into creativity matters as well. It isn't merely about minimizing harm but maximizing benefit. For instance, public art installations that double as rainwater harvesting systems provide aesthetic pleasure while contributing tangibly to urban sustainability. Through such initiatives, artists become innovators, and their works serve as practical solutions in the pursuit of conservation.

The role of music is equally transformative. Consider the global phenomenon of music festivals embracing green initiatives—from waste reduction and recycling programs to promoting alternative energy sources. These events broadcast the rhythms of sustainability to vast audiences, subtly shifting cultural norms toward greener practices.

Not to be overlooked is the educational impact of weaving sustainability themes into the suggestive power of story and song. Lyrics and narratives that speak of the Earth's fragility or the interconnectedness of all life plant seeds of ecological mindfulness in fertile ground, which grow into advocacy and action within communities.

It is evident that the creative sectors are not mere passengers on the journey to sustainability; they are driving forces. However, such roles are not always recognized. Policies and incentives that empower artists and musicians to explore and expand their participation in sustainability efforts are vital. Such support enables the creative community not only to imagine, but to actively shape a sustainable future.

Moreover, instilling conservation principles into children through creative education can provide a foundation for lifelong sustainability practices. Integrating these ideals into art and music programs in schools nurtures a generation that is inherently aware of their environmental footprint.

Nevertheless, the confluence of creativity and conservation faces challenges. The sustainability of materials and processes in art and music production requires constant vigilance and innovation. Artists must perpetually balance their creative impulses with the environmental impacts of their mediums and methods.

Creativity can also help in bridging the gap between Indigenous practices and modern conservation methods. Many traditional activities inherently possess a sustainable ethos and can provide inspiration for contemporary art forms. Doing so blends ancestral wisdom with modern sustainability pursuits.

In light of this, partnerships among conservationists, educators, artists, and musicians can yield rich dividends. By pooling knowledge and resources, these collaborations can spearhead projects that leave indelible imprints on both the environmental landscape and cultural consciousness.

However, conservation cannot trump creativity, nor the reverse. Each must inform and enrich the other. There exists a delicate equilibrium between allowing art to flourish with its boundless expression and ensuring these expressions don't contribute to planetary harm. The answer lies in fostering sustainable values within the creative process itself so both ideals can thrive in harmony.

Finally, the legacy of conservation-laden creativity is immense when viewed from a long-term perspective. Art and music that encapsulates sustainability becomes a part of cultural heritage. Through this heritage, the ideals of environmental stewardship are passed down from one generation to another in the most innate of human languages.

To conclude, the harmonization of creativity with conservation is not mere advocacy; it is an evolving canvas of human expression. It is a testament to our resilience and willingness to adapt our cultural expressions to protect what we ultimately depend on: our planet. In that, art and music don't just mirror society; they can lead it toward a greener, more conscientious, and beautifully sustainable existence.

Chapter 10:
Education as the Seed of Sustainability

A s we journey further into exploring a thriving future, we find ourselves firmly rooted in the rich soil of education, a seedbed from which the green shoots of sustainability can sprout. Knowledge empowers change, and it's within the multi-faceted education systems—from the creative classrooms that spark young minds to the ongoing training in forward-thinking businesses—where the groundwork for an enduring culture of environmental stewardship must be laid.

In this pivotal chapter, we dive into the transformative power of a multicultural curriculum that mirrors our diverse world and prepares pupils not only for their roles in society but also as caretakers of our planet. Moreover, we examine the imperatives of lifelong learning and how continuous skill-building from childhood through to the boardroom fuels innovation and instills a commitment to eco-centric business practices. These aren't merely scholarly pursuits; they are the nurturing pathways toward a sustainable ethos that will be woven into the fabric of every future generation, creating a legacy wherein each person recognizes their potential to contribute invaluably to the protection and prosperity of our global environment.

Multi-Cultural Curriculum for a Greener Tomorrow

Education, esteemed for its ability to shape minds and futures, emerges as vital when discussing sustainability. In this ideal curriculum, diversity isn't merely a checklist; it's the framework of the classroom where every lesson sown is a seed for a greener tomorrow. Within the tapestry of global sustainability, a multi-cultural approach to education infuses the knowledge and practices of various cultures into a harmonized, eco-conscious learning experience.

To instill an appreciation and understanding of sustainable practices from around the world, it's imperative that we start within the education system. Integrating a culturally diverse environmental education creates a rich soil in which an array of solutions can flourish, reflecting the world in its true heterogeneity. Here, the goal is not just to inform but to inspire actions that consider the ecological welfare of our shared planet.

Such a curriculum would harness narratives, practices, and traditional ecological knowledge from diverse societies, using these compelling stories to showcase how various communities interact with their ecosystem. As students learn about the agricultural innovations of the Ifugao rice terraces or the water conservation stratagems of the Qanat system in Iran, they are introduced to a world where sustainability is not homogenous but as varied as the cultures that fuel it.

Access to sustainability education is fundamental and should not be limited by socioeconomic or cultural boundaries. Inequality in education, especially regarding sustainability, perpetuates a cycle where those who are most affected by environmental degradation are often the least empowered to make a change. We must strive for inclusive education systems that eradicate such inequalities and offer everyone the tools to contribute to a more sustainable society.

A multi-cultural sustainability curriculum should also involve experiential learning by connecting students with local projects and global initiatives. Through these hands-on experiences, learners can witness the direct impact of sustainable practices and grasp the urgency of applying them to real-world situations.

Language and communication in a multi-cultural context must afford a bridge to convey the universal values of sustainability. Educators play the role of translators, turning lessons from diverse tongues into a common conversation on ecological consciousness. When this discourse is offered, students can articulate their own cultural identities within the larger narrative of environmental stewardship.

Moreover, the inclusion of Indigenous wisdom in sustainability education is more than an act of cultural preservation; it contributes practical methods honed over millennia to the modern sustainability toolkit. When curricula feature the rotational cultivation practices of Native Americans or the agroforestry techniques of the Kayapó, students gain respect for these deep-rooted knowledge systems and learn to value the ecological balance they ensure.

A multicultural curriculum must also examine how various societies approach sustainability differently in the face of climate change. For instance, when contrasting the resilience of social infrastructures across different cultures, students can explore how community ties affect disaster response and environmental remediation efforts.

Crucially, a sense of global citizenship is fostered through a multicultural, green curriculum. It shapes individuals who are cognizant of their impacts on environments and cultures other than their own. It's about brewing a collective responsibility that transcends borders and binds us in the common pursuit of a sustainable future.

However, integrating such diverse content into existing curricula isn't without challenges. It requires a pedagogical shift toward a systems thinking approach, one that is adept at connecting disparate cultural dots into a cohesive portrait of sustainability that students can understand and appreciate.

Teachers must also be equipped with the necessary training and resources to deliver this multi-layered content effectively. Professional development programs focused on multicultural sustainability can help educators integrate these concepts into a variety of subjects, from science and geography to social studies and literature.

Technology plays a pivotal role in this transformation, democratizing access to sustainability education. Online platforms can serve as repositories of multicultural wisdom by breaking down barriers and extending the global classroom to every corner of the planet. They can also champion collaborative learning by linking together students from diverse cultures to solve sustainability challenges.

Evaluation methods in such an educational paradigm must reflect its diversity. Assessments should not only test knowledge retention, but also the student's ability to think critically about sustainability from multiple cultural perspectives. Project-based evaluations, for example, encourage innovative thinking and practical application of the lessons learned.

The path toward a greener tomorrow is multifaceted, but one thread runs consistently through the narrative: education. A multicultural curriculum serves as a cornerstone in building a sustainable future. This signifies the importance of intercultural understanding and respect in addressing global environmental challenges. The seed of sustainability, planted in the fertile ground of informed, inclusive, and innovative education, holds the potential to sprout a future where diverse cultures thrive in ecological harmony.

Lifelong Learning: From Childhood to Business Training

In the pursuit of a sustainable future, the significance of education as a foundational element cannot be understated. Education, intrinsically linked with the concept of lifelong learning, extends far beyond the traditional confines of childhood schooling. It encompasses a continuous process, spanning from the formative years to professional development in the business world. This comprehensive approach underpins the idea that an investment in knowledge pays the best interest when addressing sustainability.

The formative years of childhood education are critical for instilling sustainable practices. Children, when exposed to multicultural environmental education, learn to respect and value the diversity of the world and its myriad ways of life. Integrating sustainability into curricula across cultures not only fosters awareness, but also empowers the youth to become proactive stewards of their environments.

As these children grow, the transition into higher education and vocational training programs offers new opportunities for advancing sustainability education. Universities and colleges around the globe are increasingly incorporating sustainability courses and degrees, often drawing on local and Indigenous knowledge to enrich the curriculum. This melding of traditional wisdom with contemporary scientific understanding advances a hybrid education that is more than the sum of its parts.

In the business sector, training and professional development have emerged as key avenues for sustainability. Companies are recognizing the value of equipping their teams with the skills necessary to innovate and implement sustainable practices. Moreover, diverse and inclusive business training can unlock the unique contributions of different cultures toward sustainability goals.

Indeed, cultural sensitivity is paramount in adult education and training. Adult learners bring rich personal and cultural histories to the table. Acknowledging this enhances the learning experience and contributes to greater engagement with sustainable practices. Lifelong learning programs must be adaptable to different cultural contexts to be truly effective.

Leadership development is another crucial aspect of lifelong learning. Leaders who are educated in sustainability principles can inspire their teams and influence their organizations to contribute positively toward the environment. Sustainability leadership courses often emphasize the importance of cross-cultural communication and the ability to mediate between different perspectives.

For professionals already in the workforce, continuous learning comes in seminars, workshops, and online courses, many geared toward sustainability. These learning platforms must reflect multinational perspectives and practices to cultivate a well-rounded understanding of global sustainability challenges and opportunities.

Technical training is also important in sustainable development, especially concerning green technologies and industries. For instance, training programs for renewable energy technicians or sustainable agriculture specialists are growing in popularity and necessity, often blending local traditional knowledge with technical skills.

The rise of digital learning platforms has greatly enhanced lifelong learning accessibility by allowing learners from various cultural backgrounds to access a wealth of information and training opportunities in sustainability that were previously out of reach. These platforms must continuously evolve to be inclusive and accessible to learners with different needs and from different cultural settings.

The role of non-formal education in engaging communities should not be overlooked. Community learning centers, local seminars, and citizen science projects also form part of the lifelong learning spectrum. They provide important community-based sustainability education, often integrating multigenerational and multicultural learning experiences that are highly localized and action-oriented.

Moreover, mentorships and apprenticeships serve as bridges between generations, allowing wisdom and skills to be handed down and adapted to contemporary sustainability challenges. These relationships foster a practical, hands-on approach to learning that deeply resonates with many cultural traditions.

The concept of lifelong learning also encourages a reflective practice where individuals and organizations take the time to assess and learn from their sustainability efforts. Reflection enables learners to understand the impacts of their actions and to plan future strategies more effectively. This cyclical process is central to continuous improvement in sustainability.

Critical thinking and problem-solving skills—key lifelong learning outcomes—are essential for tackling the complex issues underpinning sustainability. Education systems need to nurture these skills from an early age and retain a focus on them throughout higher education and professional development programs.

Furthermore, ethical considerations play a pivotal role in sustainability training across all age groups and professional levels. Ethics-based education helps individuals and organizations make responsible decisions that reflect not only economic and environmental considerations, but also social justice and equity.

In conclusion, lifelong learning—from childhood through to business training—is a vehicle for empowering individuals and communities to be active participants in the drive toward sustainability. It bridges gaps across cultures and generations and fosters a shared responsibility for creating a resilient and sustainable world. Shaping capable, knowledgeable, and ethical minds is essential for overcoming the myriad environmental challenges of our time and for advancing global sustainability well into the future.

Chapter 11:
Green Economics: Weaving
Prosperity with Planet

A s we transition into an era that recognizes the finite nature of our resources, a robust conversation blossoms around the fusion of economy and ecology. This chapter delves into the transformative concept of green economics, an approach that doesn't merely tolerate environmental consideration as an add-on, but elevates it to the cornerstone of economic prosperity.

Here, prosperity is redefined, not by the relentless pursuit of growth at the expense of the planet, but as a balanced journey toward well-being and abundance that honors the Earth's ecosystems. This symbiosis is achieved by incentivizing businesses to adopt sustainable practices and demonstrating that long-term profitability aligns with ecological stewardship. The chapter illustrates the emergence of innovative business models that strike a harmony between cultural value systems and environmental imperatives, highlighting the potential for culturally aware green economies to create profitable and sustainable opportunities. It isn't a tale of trade-offs but rather an inspiring narrative of synergies and the idea that the future echoes with the promise of economies that thrive in concert with the planet.

Business Models that Integrate a Multi-Cultural Approach to Sustainability

The concept of sustainability transcends borders, cultures, and economies. It is anchored in the recognition that a diverse yet interconnected planet has to find unified ways to thrive. Green economics—an emerging paradigm—excels when it incorporates a multi-cultural lens. In this section, we examine business models that do not simply graft sustainability practices onto existing frameworks but are deeply rooted in multi-cultural ideologies as well.

The success of these business models lies in their ability to intertwine economic viability with cultural sensitivity and ecological responsibility. A company that wishes to lead in the green economy cannot afford to be myopic regarding cultural narratives and practices. Instead, it must adapt and innovate by observing and learning from various cultures and their relationship with the environment.

Consider a business that sources its materials from Indigenous suppliers. It enters into a partnership that's not only economic, but also cultural and ecological. It knows sustainable practices are ingrained in many Indigenous cultures, and by learning from and with these communities, the business can ensure that its operations support biodiversity, conservation, and social equity.

Moreover, these business models leverage the diverse cultural approaches to sustainability to access a richer array of solutions to environmental challenges. They recognize a practice deemed sustainable in one cultural context may not work as effectively in another, prompting a more innovative and adaptive approach.

Multiplicity in approaches also enables businesses to navigate the varied regulatory landscapes that exist worldwide. For instance, a business model adaptive to local traditions will more effectively meet or exceed environmental regulations in different countries by avoiding the one-size-fits-all pitfall of some global sustainability standards.

Another hallmark of these enlightened business models is their emphasis on building long-term relationships with communities. This alliance goes far beyond transactional interactions and delves into reciprocal partnerships. They are also embodiments of corporate social responsibility, where success is not solely about financial gains, but also about improving the quality of life for people and nurturing the planet.

For global businesses, the multicultural approach to sustainability also entails linguistic and educational diversity. Companies train their employees in cultural competence, understanding that a monolingual and monocultural workforce will likely miss opportunities in the global market. This cultural agility not only fosters inclusivity, but also drives innovation and creativity.

Marketing strategies within these business models must also be culturally cognizant. Products and services are presented not just as "green" or "eco-friendly," but in ways that resonate with the local ethos. For instance, in regions where water is considered sacred, water-saving products are marketed not just for their cost-saving benefits, but as tools to honor a revered resource.

Further, these business models pay close attention to the product life cycle in a multi-cultural context. From design and production to distribution and disposal, each phase is examined through a cultural lens to ensure it aligns with the environmentally sustainable practices of various cultures.

The digital realm has also fostered an unprecedented platform for integrating multi-cultural approaches to sustainability. Businesses leverage technology to share, learn, and collaborate across borders. However, they also recognize the importance of digital equity and aim to ensure the technological advantages they employ do not widen the digital divide.

More fundamentally, businesses that take multiculturalism seriously in their sustainability approaches often advocate for systemic change. They understand it's not just about creating green products and services, but also about participating in the larger conversation and movement toward a sustainable global economy.

In conclusion, pioneering businesses in the green economy are those that take a worldly view, value diverse cultural inputs, and honor the complex fabric of global ecosystems. They seek a collaborative approach that fuses traditional wisdom with modern innovation and benefits not just their bottom line, but also contributes to the collective well-being of communities and the planet. They hold themselves accountable to the highest standards of ethical practice and ecological stewardship—because the future of our world depends on it.

Cultural Capital in the Green Economy

As we delve into the crescendoing symphony of the green economy, it's essential to acknowledge the subtle yet potent role of cultural capital. The heritage, knowledge, and skills unique to cultures worldwide foster a rich soil from which sustainable practices can flourish. Cultural capital in the green economy not only amplifies biodiversity and conservation efforts, but also buttresses economic resilience and innovation by tying the threads of cultural diversity into the broader fabric of global sustainability.

The concept of cultural capital—the non-financial social assets that promote social mobility—finds fertile ground in green economics. Within Indigenous communities, age-old knowledge about land management practices, traditional crops, and natural remedies represents invaluable assets. A green economy thrives when such Indigenous wisdom is not merely preserved but actively integrated into sustainable development, tourism, and educational initiatives.

Moreover, the green economy doesn't stand in isolation; it's deeply enmeshed with the varied ways communities have traditionally interacted with their environment. Agricultural practices honed over millennia, such as crop rotation and organic farming, carry lessons on sustainable yield and soil health that are vital today. They embed an ecological intelligence into economic systems that modern industries often bypass, to their detriment.

Across oceans and landmasses, artisans of the world have cultivated a relationship with their local environments, extracting materials and inspiration to fuel economies of craftsmanship. This sector of the green economy overlays cultural capital onto sustainable production, making each creation a story of the ecosystem from which it emerged. Societies valuing such craftsmanship are naturally inclined toward supporting local economies, reducing the carbon footprint through minimal transportation, and advocating for fair labor practices.

Fostering cultural capital is not without its challenges. Globalization, while offering an expansive market for goods and services, can erode the uniqueness of local cultural identities. In a competitive global market, the temptation to conform and adopt homogeneous business models can stifle cultural diversity. In contrast, enterprises that leverage their distinct cultural heritages forge a unique identity and foster a brand that's not only internationally recognized, but also locally cherished.

Consider the example of eco-tourism, where cultural capital is the cornerstone of an industry that balances profitability with the preservation of natural and cultural heritage. Eco-tourism invites visitors to immerse themselves in local traditions and ecosystems, which in turn educates the visitors and generates revenue to maintain those very environments and cultures. This showcases the value of cultural capital in guiding economic models to honor and safeguard the planet's diversity.

The green economy also benefits from the transmission of cultural capital through education and mentorship. Business incubators and educational institutions nurturing eco-entrepreneurship should incorporate cultural narratives and practices into their curricula. This fusion empowers new visionaries to create business solutions grounded in cultural understanding and sustainability.

Moreover, the intersection of culture and the green economy offers an avenue to address injustices against marginalized communities. By valorizing the cultural capital of these groups, green economies can counteract the legacy of extraction and exploitation with models that promote equity and inclusion. Fair trade and direct trade models in agriculture, for example, are built on the premise that the custodians of the lands from which products originate should receive fair compensation for both their commodities and their knowledge.

As urbanization continues, the cultural capital of rural communities risks being overlooked. Yet, the preservation of this capital is essential for fostering heterogeneity within the green economy. Urban farming initiatives and community gardens, for instance, can serve as hubs that not only provide local produce, but also revitalize cultural practices linked with agriculture in urban settings.

Cultural festivals, too, harness the potential of cultural capital for the green economy. They become platforms for ecological messaging by showcasing sustainable lifestyles and facilitating the exchange of green technologies hand in hand with cultural celebration. They knit the community closer, emphasizing a collective identity that's vested in both cultural preservation and environmental stewardship.

Investing in cultural capital transcends economic returns. It hedges against monocultures of the mind and the land, which are narrowly defined frameworks that fail to capture the nuances of ecosystems and human imagination alike. When adequately leveraged, cultural capital builds an economy that's vibrant, adaptive, and rooted in the collective heritage of its participants.

Negotiating intellectual property rights within the green economy is a crucial aspect of preserving cultural capital. Artists, scientists, and traditional knowledge holders must navigate a complex landscape where their cultural contributions are protected while shared for the greater good. Here, implementing benefit-sharing agreements can ensure communities are fairly compensated and acknowledged for their contributions to biodiversity and knowledge-sharing platforms.

Within the policy realm, cultural capital must be given a seat at the table. Policymakers have the authority to create frameworks that elevate and integrate cultural practices into sustainable economic strategies.

These policies should aim to nourish the green economy while protecting and promoting the intangible assets that endow it with character, resilience, and depth.

In summary, cultural capital is not an ancillary feature of the green economy; it is its beating heart. It cultivates a fertile ground where prosperity and planet are not separate aspirations but part of a singular, thriving ecology. The green economy, therefore, must be an ever-evolving tapestry, interwoven with the vibrant threads of cultural capital, creating a tableau rich with diversity, wisdom, and sustainable strength.

Chapter 12:
Rural Traditions and Innovations

I n the rich tapestry of rural life, tradition and innovation intertwine to form resilient threads that hold the potential for sustainable progress. This chapter delves into the remarkable ways rural cultures preserve their heritage while simultaneously embracing new methods in agriculture and community building. It spotlights the unsung heroes who uphold time-honored practices that have been the cornerstone of ecological balance and who boldly innovate, ensuring that tradition evolves in harmony with the needs of a changing planet.

The chapter reveals how ancient wisdom, when fused with modern technology and sustainable practices, not only sustains but revitalizes rural communities by preserving their cultural identity and encouraging a rebirth of environmental stewardship. Within this narrative, we'll uncover the resilience rooted in agricultural practices that shield and enrich the Earth, serving as a beacon of sustainability that can light the way for urban and global counterparts.

As we explore the symbiotic relationship between old and new, we'll see that the heart of rural innovation lies in its people's ability to adapt— ensuring food security, biodiversity, and the conservation of natural resources for generations to come.

Preserving and Reinventing Rural Cultures

Rural cultures around the world are reservoirs of profound knowledge, traditions, and practices that have been refined over millennia. These cultural expressions evolved in response to the landscapes in which

106

communities live and are a tangible manifestation of a people's connection to their environment. However, with the rapid changes that techno-industrial society has imposed on the planet, safeguarding these traditions while encouraging innovation poses a critical challenge. This section delves into how rural traditions can be both preserved and revitalized to advance sustainable living.

Remote rural communities often exemplify a deep understanding of their local ecosystems. For instance, some have developed intricate agroforestry practices that maintain both forest cover and agricultural productivity. These practices challenge modern norms of monoculture and offer valuable lessons on biodiversity conservation and sustainability. They represent an intact cultural fabric that operates in full cognizance of nature's intricate systems and is attuned to the subtle signals of the Earth.

In preserving rural cultures, it is essential to recognize the sovereignty of local knowledge systems. Traditional ecological knowledge, which encompasses spiritual relationships, practices, and rituals, requires not just respect but legal and societal acknowledgment. Policies that support traditional practices in land management, for example, can contribute to resilience against climate change while asserting the right of Indigenous peoples to their ancestral lands.

However, the mere preservation of culture is not sufficient. Rural communities, particularly those in very remote areas, often suffer from a lack of infrastructure, services, and opportunities. This can lead to outmigration, particularly of the youth, and a consequent erosion of culture. Reinventing rural cultures in a manner that is true to their roots yet appeals to younger generations is critical for their continuation.

Digital technology, often associated with urban environments, can play an unexpected role in preserving rural cultures. It helps maintain cultural practices across dispersed populations and provides platforms for sharing stories, music, language, and art . Furthermore, technology can support rural economies and serve as a bridge between traditional knowledge and modern science, generating new, more sustainable solutions to contemporary problems.

In the most distant communities, isolation can be both a protective barrier and a substantial hurdle. While such isolation can safeguard cultures from external influences that dilute traditional practices, it can also hinder the integration of these valuable cultural elements into broader sustainability discussions. Here, maintaining a delicate balance between exposure and protection is crucial to ensure both the survival and evolution of these cultures.

Renewable energy presents an excellent example of a modern innovation that can serve rural cultures. Innovation can meet energy needs while respecting the integrity of the landscape. Solar, wind, and small-scale hydroelectric projects that communities control and manage themselves can power schools, health centers, and communications infrastructure. This kind of progress respecting autonomy and tradition illustrates the complementary nature of innovation and tradition.

Food sovereignty is another realm in which rural cultures thrive. Time-honored agricultural practices that are inherently organic and non-GMO can become part of the global discourse against the industrial food complex. By creating market opportunities for local, traditionally farmed products, rural communities can introduce the world to sustainable agricultural practices that also preserve cultural heritage.

The arts and oral traditions, which are often central to rural communities, possess immense power to convey messages of sustainability and resilience. The narrative arts—such as storytelling, song, and dance—communicate values and knowledge through generations. These practices can be repurposed in service of sustainability education, demonstrating that cultural preservation and innovation are not mutually exclusive.

Community governance structures often embody principles of sustainability that are lost in larger, more bureaucratic systems. Decision-making that respects elders and community assemblies underlines the significance of collective interests over individual gains. In many rural societies, decisions are made with far-sightedness, considering the impact on seven future generations, a concept that significantly resonates with sustainable development goals.

It is, however, essential to facilitate within communities dialogues about what aspects of culture they wish to preserve and what innovations they seek to embrace. Imposing a vision of cultural preservation from the outside can be seen as neo-colonial and may fail to serve the community's actual needs and desires.

Understanding this delicate balance between preservation and innovation sets the stage for appreciating the significance of cultural festivals and rituals. These events, often tied to agricultural cycles or natural phenomena, underscore the deep connection between community life and the surrounding ecosystem. By celebrating these traditions, communities not only reaffirm their cultural identity but also open doors to sustainable tourism. This approach not only provides a source of income for rural areas but also plays a crucial role in educating visitors about the value of protecting both cultural and environmental treasures.

Finally, in considering the preservation and reinvention of rural cultures, the education system plays a foundational role. Culturally relevant curricula that include local languages, histories, and sciences foster a sense of pride and interest in one's heritage among the youth. Such education must be a collaborative effort with local leaders and elders to reflect and transmit the wisdom of rural cultures accurately.

As we consider the future of sustainability, we must embrace the richness and diversity of rural traditions as part of our global heritage. It's not merely an act of conservation; it's recognizing these cultures carry vital seeds for the future of human resilience and our planet's health. By protecting these seeds and allowing them to germinate through the careful application of innovation, rural cultures can continue to be wellsprings of sustainability.

Throughout this discourse, it must be remembered that preserving and reinventing rural cultures is not just an academic exercise but a day-to-day, lived experience for millions of people. It is only through their continued existence and dynamic adaption that we can aim to craft a truly sustainable and resilient world.

Agriculture: The Root of Resilience

In the vast expanse of rural traditions, agriculture stands as a testament to human ingenuity and resilience. It's a thread that connects generations by weaving together lessons from the past with innovations for the future. For centuries, farmers have cultivated the land, inherently understanding the health of their crops and their communities depends on the health of the Earth itself. Such recognition is at the heart of regenerative agriculture—a concept that merges ancient wisdom with modern practices to create sustainable food systems.

Regenerative agriculture is not just a method; it's a philosophy that honors the symbiosis between land and cultivator. This approach harks back to Traditional Ecological Knowledge (TEK), which encapsulates the understanding Indigenous and local communities have of their environment. These knowledge systems, rooted in long-term observations and experiences, emphasize respect for natural cycles and promote practices that support ecosystem health.

At its core, regenerative agriculture seeks to renew and build soil health while also addressing issues like climate change and biodiversity loss. The principle of disturbing the soil as little as possible, for instance, preserves the complex network of organisms that is vital to soil fertility—a lesson learned from observing natural ecosystems. Cover cropping, another regenerative practice, mimics the uninterrupted plant cover of a natural landscape, which protects and nourishes the soil.

Moreover, crop rotation and diversity, staples of regenerative agriculture, have been utilized for millennia as farmers understood the dangers of monocultures long before modern science did. These practices mirror natural ecological processes, where diversity equates to resilience. By rotating crops and embracing polycultures, farmers can naturally disrupt pest cycles and improve the land's resilience to climatic stressors.

The resilience offered by regenerative practices also extends to water management. Techniques such as swales and rain gardens enhance the land's water retention capabilities, which can be life-saving in areas facing water scarcity. This is especially crucial in rural regions where agriculture is often the linchpin of the local economy and community well-being.

The role of animals is also crucial to regenerative agriculture. Integrating livestock mimics the way wild animals interact with the land in natural ecosystems. Managed properly, grazing can help in nutrient recycling, soil aeration, and seed dispersal, contributing to a healthier agroecosystem.

However, regenerative agriculture is about more than just looking back. It's about using our historical and cultural knowledge as a springboard for innovation. Modern technology and scientific understanding enhance TEK, guiding us toward new solutions rooted in age-old wisdom. Data-driven practices can tell us more about the intricate workings of soil microbiomes, helping refine regenerative techniques for even better outcomes.

The challenges of modern agriculture—degraded soils, water shortages, and the pressing need to feed a growing global population—make the role of regenerative practices more prominent than ever. These issues call for a holistic perspective that acknowledges the interconnectedness of human activities and natural systems.

Agriculture's role in building resilient rural communities can't be overstated. Local food systems based on regenerative practices are more than just a buffer against the unpredictability of global food chains; they are a means of preserving cultural heritage while progressing toward sustainability. They imbue the community with a sense of autonomy and empower individuals through the knowledge that their actions directly contribute to environmental stewardship.

Empowered rural communities are better positioned to face the impacts of climate change. Farming methods that improve soil carbon sequestration, for example, turn agriculture from a climate change contributor into a potential solution. As these communities adapt to changing conditions, they create a blueprint of resilience that can inform broader efforts to mitigate and adapt to climate change.

Furthermore, embracing regenerative agriculture aligns with efforts to build a greener economy. The methodology complements the push for sustainable development by drawing a clear line between the health of the environment and economic prosperity. By fostering ecosystems that are more productive and resilient, regenerative practices enhance the economic stability of rural areas, reducing the reliance on external inputs and the vulnerabilities they create.

Education plays a pivotal role as well, bridging generations and geographies. Sharing knowledge about regenerative practices ensures farmers, young and old, have the tools to adapt to an evolving world while maintaining their cultural identity. This fusion of education, tradition, and innovation is vital for sustaining rural communities in the face of modern challenges.

Partnerships, too, are instrumental in scaling regenerative agriculture. Collaborations among farmers, scientists, and policymakers can lead to supportive frameworks that promote the spread of regenerative practices. By creating incentives and removing barriers, these partnerships can transform the patchwork of regenerative farms into a quilt of resilience stretching across the rural landscape.

Conclusively, agriculture isn't just the root of rural resilience; it's a branch reaching into the future and offering hope. It signifies a commitment to life-giving practices that honor the past while nurturing the promise of tomorrow. As society becomes ever more aware of the need for sustainable approaches, the lessons from rural traditions and regenerative agriculture will be crucial guideposts on the journey to resilience.

Chapter 13:
The Fabric of Social Sustainability

I n the tapestry of global sustainability, the social dimension forms a crucial thread that is integral to holding together the pattern. This chapter delves into the intricate texture of social equity and inclusion, recognizing that environmental concerns can't be separated from social justice issues. It weaves through the multifaceted nature of community well-being, identifying cultural approaches to health that underscore the importance of recognizing diverse needs and practices in fostering overall societal health.

As we explore the interpersonal strands that crisscross through communities, we uncover the foundational belief that sustainability is not only an ecological or economic challenge but also a deeply social— human—one. It's about ensuring no one is left behind in pursuing a resilient and vibrant society. This imperative obliges us to confront and dismantle historic injustices and move toward systems that enable participation and benefit sharing across all societal cross-sections. Hence, in sculpting the fabric of social sustainability, we draw from a rich palette of community wisdom, integrating inclusion and wellness into the very core of sustainable development.

Social Equity and Inclusion

As we peel back the layers of social sustainability, the core reveals itself to be entrenched in social equity and inclusion. These principles are not merely aspirational; they represent foundational members in the architecture of a society that aspires to be truly sustainable. This entails providing every individual, regardless of their background, with fair access to resources, opportunities, and support so they can contribute to and benefit from a sustainable future.

At the heart of social equity is the recognition that societies thrive when diversity is not just acknowledged but celebrated and integrated. This acceptance can manifest through policies that promote inclusion in educational settings, workplaces, and governance. Such initiatives forge pathways for marginalized groups to influence decisions that affect their lives, allowing for a multiplicity of perspectives at the decision-making table.

Historically, many communities have been left on the peripheries of development processes, leading to deep-rooted inequalities and societal fault lines. Bringing these groups into the fold isn't just a moral imperative; it also makes practical sense as it taps into a wellspring of untapped potential. Inclusion breeds innovation and ensures a wider array of solutions to environmental and social challenges.

Inclusivity does not imply a one-size-fits-all approach. Instead, it demands contextual sensitivity and a willingness to adapt strategies to fit diverse cultural, social, and economic landscapes. Sustainable development initiatives can falter if they fail to resonate with the values and realities of the people they are intended to serve. Therefore, it is crucial to engage in meaningful dialogues with individuals from various societal sectors to ensure sustainability is not something that is done to them, but something in which they actively participate.

Education plays a pivotal role in uplifting communities and fostering social inclusion. When individuals are knowledgeable about sustainability and their rights within this domain, they become empowered to demand better, fairer systems that reflect their needs. Furthermore, inclusivity in education ensures a broader array of ideas, viewpoints, and innovations that can drive sustainable development.

Ensuring gender equity is another crucial dimension of social equity and inclusion. The unique challenges women and other gender minorities face often go unaddressed, particularly in the context of climate change and environmental degradation. Empowering these groups not only drives progress toward equality, but also harnesses their invaluable insight and experience in crafting sustainable solutions.

Income inequality is another barrier to social sustainability. The widening gap between the rich and the poor can lead to social unrest, decreased economic stability, and diminished health outcomes for vulnerable populations. By bolstering the financial resilience of low-income families through equitable access to sustainable jobs and resources, society at large can reap the benefits of a more cohesive and cooperative populace.

Inclusive healthcare is yet another cornerstone of a socially sustainable future. Disparities in health accessibility and outcomes can perpetuate cycles of poverty and marginalization, whereas equitable health systems that recognize and accommodate cultural nuances can improve community resilience.

Urban planning and the development of inclusive cities are integral to fostering social equity. When urban environments are designed with all citizens in mind, including the most marginalized, the result is safer, healthier, and more sustainable communities. These spaces empower individuals by promoting a sense of belonging and enabling a more dynamic interaction among different societal groups.

Access to green spaces and environmental resources is equally critical. Regardless of socioeconomic status, everyone deserves the opportunity to experience and benefit from nature. Policies that strive to even the distribution of environmental goods, such as clean air and water, green parks, and toxin-free environments, are fundamental to ensuring sustainability is a shared benefit, not a luxury.

Addressing systemic biases and fostering diversity in leadership roles reflect a society's commitment to social equity. Bringing varied perspectives to the forefront can inform policies and strategies that are more representative of underserved populations. In this sense, inclusion allows for a richer, more nuanced understanding of the complexities of sustainability challenges.

Criminal justice reform is another area where social equity plays a pivotal role. A sustainable society cannot thrive if it systematically disenfranchises portions of its population. Reforms aimed at addressing biases, providing restorative justice, and creating opportunities for reintegration underline the principles of a socially equitable system.

The value of inclusivity of cultural practices and traditions within the larger sustainability dialogues can't be overstated. Respecting and preserving cultural diversity while integrating traditional knowledge systems fosters a holistic approach to sustainable practices.

In conclusion, social equity and inclusion are far from being peripheral concerns; they are central to the fabric of social sustainability. They are the warp and weft that strengthens society, making it resilient and adaptive. As sustainability becomes increasingly mainstream in our collective consciousness, equity and inclusion will determine not only if we can achieve our goals, but also the kind of world we leave behind.

The Weft of Wellness: Cultural Approaches to Health

In our ongoing exploration of the tapestry that constitutes social sustainability, it becomes imperative to address one of its most critical strands: health. Health, much like the very weft threads in the fabric, interlaces the structure of social well-being, holding it together with resilience and durability. Just as diverse patterns enrich a tapestry, various cultural interpretations and practices enrich our understanding of wellness.

Take the interconnectedness of community health in remote, tribal societies, for instance. Here, traditional knowledge of medicinal plants passed down through generations is not only a resource for healing, but also a bond that ties the community to its environment. Far from the clinical corridors of modern hospitals, these often-small, isolated communities rely on rich pharmacopeia sourced directly from their natural surroundings.

In the Amazonian Basin, for example, the intimate knowledge Indigenous tribes hold about the rainforest is an embodiment of health that extends beyond the individual to encompass the entire ecosystem. In these rainforests, the Yanomami understand the health of their people and the forest are inseparable. Their approach to health includes plant-based remedies as well as a holistic worldview that upholds the sanctity of their environment.

Similarly, in the highlands of Papua New Guinea, the concept of *kastom*, which refers to traditional practices and social norms, includes a holistic approach to wellness that incorporates not only physical, but also spiritual and communal health. The balance and harmony that sustain the health of the community are mirrored in the rituals, ceremonies, and medicinal knowledge that govern their way of life.

Contrary to the reductionist model of health often found in industrialized contexts, Indigenous and remote communities usually perceive health in a more holistic scope. For them, wellness weaves together the physical, mental, emotional, and spiritual threads of life. In the sparsely populated Tundra regions, the Nenets herders treat health as a continuum of survival and adaptation. This has resulted in a people remarkably resilient to the harsh conditions and equipped with knowledge about decade-tested dietary practices and the healing properties of the Arctic flora.

These varied cultural practices offer invaluable insights into sustainability. They showcase that true wellness extends beyond hospital walls or pharmaceutical interventions. It envelops the environment, diet, community support, and spiritual fulfillment. This is particularly evident in how some remote Himalayan villages maintain health. Their close-knit social structures provide a network of care and support for the sick, while their traditional diets, which are rich in locally sourced grains and greens, promote longevity and natural balance.

Moreover, in such societies, the transmission of health knowledge is often a sacred duty, ensuring the wisdom of ancestors continues to nurture future generations. The Maasai of East Africa, known for their deep-rooted pastoral lifestyle, carry the legacy of their nomadic ancestors through oral tradition. Elders bequeath knowledge about the nutritional and medicinal qualities of countless herbs and plants to the young.

The resilience of these practices in the face of globalization and the pressing tide of modernization underscores the vitality that cultural approaches can offer to the concept of wellness. They teach us that sustainability is inherently a community endeavor, where the health of one reflects the health of all. Even amid global pandemics, these inherent cultural strengths reinforce the importance of unity and the communal pursuit of well-being.

In considering the weft of wellness, it's important to recognize the risk traditional health knowledge faces from the biodiversity loss, climate change impacts, and cultural erosion. As globalization encroaches on these remote communities, the very fabric of their approach to health is threatened. This has the potential to sever the passing of generational knowledge and the sustainable use of local medicinal resources.

Therefore, it becomes essential to safeguard this knowledge just as we work to protect endangered species and ecosystems. Initiatives like the Traditional Healers and Modern Practitioners programs in regions of Africa and South America aim to bridge the gap between traditional and modern medicine, emphasizing the intercultural understanding and respect that is crucial for the continuation of these practices.

As we paint this picture of diverse health tapestries, we're not just advocating for the preservation of ancient wisdom out of respect for culture. Adopting such practices within the broader framework of global health could alleviate some of the pressures on healthcare systems while offering sustainable and preventive approaches to wellness. By integrating these models into modern healthcare, the West can learn valuable lessons about breeding resiliency against chronic diseases, mental health issues, and other ailments exacerbated by modern living conditions.

When we acknowledge the tapestry of cultural approaches to health, we can see there is no singular fabric of well-being. Each community's unique pattern deserves its own study and, where beneficial, integration into the quilt of global health practices. As we consider future steps in advancing social sustainability, let's reflect on the wisdom each thread of cultural knowledge brings to the overall pattern of human wellness. By interweaving these cultural threads, we can strengthen the weft of global health and, consequently, the larger fabric of social sustainability.

Chapter 14:
Policy Patterns for a Sustainable Future

I
n "Policy Patterns for a Sustainable Future," we delve into the complex tapestry of environmental regulation threaded with cultural sensitivity, continuing the diligent efforts to unite humanity's rich diversity with effective sustainability strategies. This chapter explores the profound necessity of tailoring policy interventions to honor the intricate blend of cultural values and environmental imperatives. Crafting informed policies that are sensitive to cultural nuances requires not just a deep understanding of environmental science, but also a respect for the medley of global traditions that influence how communities interact with their natural surroundings.

We highlight the significance of cultural sensitivity in environmental regulation as a keystone for engendering cooperative, successful, and just policies. Multilateral cooperation is essential, and we unravel how policy embroidery—the intricate interweaving of agreements and treaties at multiple levels—can stitch together a resilient global response that respects local traditions while addressing universal challenges. Through a fusion of scientific preciseness, persuasive narratives, and motivational examples, we envision a future where policies align with both ecological wisdom and the endurance of cultural legacies, fostering a sustainable path forward that is as diverse and dynamic as the people it is designed to protect.

Cultural Sensitivity in Environmental Regulation

As we step into the discourse of cultural sensitivity within environmental regulation, it's clear that acknowledging and respecting the diverse expressions of humanity's relationship with the natural world is not only a matter of principle, but also a strategic imperative. Regulatory frameworks that are attuned to cultural diversity are a cornerstone in the endeavor to forge a sustainable future. Environmental justice, as an ethical and political issue, demands we consider the unique cultural, historical, and socioeconomic contexts of communities such policies affect.

The successful implementation of environmental regulations often hinges on their cultural appropriateness. When policymakers overlook the cultural dimensions of the communities they intend to serve, they run the risk of creating resistance and inefficacy. This misalignment stifles proactive environmental stewardship and may exacerbate existing inequalities, especially among marginalized communities.

Environmental justice encompasses the fair treatment and meaningful involvement of all people—regardless of race, color, national origin, or income—concerning the development, implementation, and enforcement of environmental laws, regulations, and policies. This principle is integral to culturally sensitive environmental regulation. It requires us to consciously design policies that avoid the imposition of harmful environmental conditions on disadvantaged groups while distributing environmental benefits equitably.

Recognizing the right of all people to a healthy environment also means acknowledging the intimate connection many cultures have with their ancestral lands. This connection is not just spiritual or traditional but often empirical, reflecting centuries of sustainable living and environmental management. It's a knowledge base that can greatly enhance modern environmental regulation, provided it's integrated respectfully and collaboratively.

Efforts to introduce culturally sensitive environmental regulation can come in many forms. One of the most powerful is the inclusion of Indigenous and local communities in decision-making processes. This approach moves beyond consultation to genuine co-management, reflecting the notion that those most affected by environmental policies should have a significant voice in crafting them.

The concept of Free, Prior, and Informed Consent (FPIC) has emerged as a benchmark for respecting Indigenous rights in the context of environmental regulation. FPIC implies that communities have the opportunity to approve or reject projects that may affect their lands and livelihoods before any operations commence. Embracing FPIC is a critical component of empowering communities and recognizing their sovereignty.

Cultural impact assessments (CIAs) offer a mechanism for evaluating how proposed environmental regulations might affect the cultural resources and practices of a community. CIAs provide valuable insights that can lead to regulations that are not only environmentally sound, but also culturally congruent, thereby fostering greater compliance and support from the community.

In the context of climate change, certain cultural practices can offer adaptive strategies that enhance resilience. For example, traditional agricultural methods and local knowledge of ecological cycles can be integrated into climate change adaptation policies to increase the effectiveness and acceptability of these policies in the local setting.

This integration, however, should not be superficial or tokenistic. Genuine cultural sensitivity means enabling meaningful participation, where the knowledge and values of all stakeholders are weighed and respected. This can entail overcoming language barriers, alternative legal recognition, and the creation of forums in which different knowledge systems coexist and enrich each other.

As we venture to instill cultural sensitivity into environmental regulation, it's crucial to address the contested nature of "cultural values." Dynamic and constantly evolving, culture defies static definitions. Laws and policies must be flexible enough to adapt to changing cultural scenarios and circumstances, thus ensuring their continued relevance and fairness over time.

Moreover, culturally sensitive environmental regulation ties directly into the broader aims of sustainable development. It aligns with the United Nations Sustainable Development Goals (SDGs), particularly those related to reducing inequalities (Goal 10), ensuring inclusive and equitable quality education (Goal 4), and fostering peace, justice, and strong institutions (Goal 16), showing how cultural sensitivity in environmental law reinforces global sustainability commitments.

When legislation acknowledges the varying cultural dimensions of sustainability, it encourages stewardship and ownership of environmental initiatives. A prime example is the regulatory protection of sacred natural sites, which simultaneously conserves biodiversity and upholds cultural rights, thereby weaving a double helix of ecological and cultural integrity.

Finally, continuous learning and adaptation are imperative to maintain cultural sensitivity in environmental regulation. No one size fits all; each community holds its own values, practices, and connections to the environment that must be recognized. This understanding should be a perpetual commitment to reflection, dialogue, and improvement as the cultural landscapes within which regulations operate continue to shift.

As we journey forward, we must remain mindful of the rich tapestry of humanity's diverse cultures. Integrating cultural sensitivity into environmental regulation is not just an ethical necessity, but also contributes to the robustness, effectiveness, and justice of our sustainability strategies. It's about creating policies that protect our planet while honoring and leveraging the vast array of human experiences, wisdom, and perspectives.

Multilateral Cooperation and Policy Embroidery

Moving through the intricacies of global sustainability, one finds the interweaving of policies and practices akin to an elaborate tapestry, each thread reliant on another. This confluence of ideas and actions is central to multilateral cooperation, where countries, organizations, and sectors unite, guiding the intricate pattern-making required for a sustainable future. Here, policy embroidery refers to the finesse with which these different entities harmonize their strategies to enhance the integrity and beauty of our shared global environment.

In the sphere of sustainability, multilateral cooperation is not a mere alliance of convenience but a profound commitment to a common vision. This is particularly evident in international agreements like the Paris Agreement, where consensus on climate goals demonstrates the possibility of cohesive global action. Countries each contribute their stitch to the tapestry, holding the fabric of our planet together through shared responsibility.

The effectiveness of policy embroidery can be seen in the synergy these cooperative efforts create. Action on climate change, for example, necessitates a fusion of economic, environmental, and social policies. Intertwining these threads of policy across borders creates a resilient and adaptable fabric that can sustain the weight of abrupt changes and slow transformation alike.

A key element of successful multilateral cooperation is the recognition and value of diverse contributions. Much like the varying textures and colors in an embroidered piece, the unique policy inputs from different cultures and economies enrich the overall pattern. This diversity brings about innovation, which allows for a range of solutions to emerge, each tailored to address specific challenges, yet contributing to the collective goal.

Another crucial aspect is policy coherence, where disparate actions come together in a cohesive strategy. It requires meticulous coordination to ensure each policy is not an isolated patch but part of an interconnected design. Such coherence amplifies the impact of sustainability measures, streamlining efforts across international, national, and local levels.

Operating within a framework of shared values and principles fortifies the fabric of multilateral cooperation. It lays the foundation upon which trust is built, making it possible for nations to work together despite varying interests and developmental stages. Common but differentiated responsibilities, a principle enshrined in international environmental law, exemplify this approach by acknowledging inequalities while fostering collective progress.

Financial mechanisms play a cardinal role in reinforcing the multilateral tapestry. The provision of funds to support climate adaptation and mitigation in developing countries, for instance, is a clear demonstration of solidarity and common purpose. The creation of the Green Climate Fund illustrates how investment threads can support the larger sustainability aim by assisting those who are most vulnerable to environmental changes.

Policy flexibility is also crucial as it allows the tapestry to evolve as circumstances change. Adaptive governance enables policies to be responsive, much like an artist adjusting their technique in reaction to the medium's response. This flexibility is necessary as we learn more about the systems we seek to sustain and the impacts of our interventions.

Engagement at all levels deepens the patterns of sustainability policies by ensuring local perspectives inform international agendas. This alignment can amplify the effectiveness of environmental actions, anchoring them in the lived experience of communities around the world. In this way, policies are not only designed from the top down but are crafted with insight and expertise drawn from grassroots levels.

Transparency and accountability are akin to the backstitches in embroidery, which hold the fabric together even when stretched. They are critical components that maintain the integrity of multilateral efforts, providing clear reporting and review systems to assess progress and guide corrections. These practices ensure commitments are met and that the collective endeavor maintains its direction and purpose.

Education and public awareness weave knowledge into the shared tapestry, empowering individuals and communities to act. They form integral patterns within policy frameworks, as informed citizens are better equipped to advocate for sustainable practices and hold their governments accountable. Education on sustainability equips the weavers—every global citizen—with the skills to contribute to this grand design.

Lastly, the resilience of this policy embroidery relies on the strength of its weakest links. International cooperation must extend support to bolster these frays, whether they are found in the smallest island nations or the most economically disadvantaged communities. Such solidarity ensures the collective work does not unravel, but instead forms a more durable and equitable fabric for the world.

As this chapter elucidates, the policy patterns for a sustainable future can be imagined as a form of artistry, a delicate yet deliberate process of bringing together disparate threads to create something that is both beautiful and vital. Crafting a sustainable future through multilateral cooperation and policy embroidery is a testament to humanity's ability to unite in the face of shared challenges. Like any masterpiece, it requires vision, dedication, and a recognition of the interconnectedness of every thread in the grand tapestry of life on Earth.

Chapter 15:
Telling the Stories of Climate Change

As we delve into the potent narratives of climate change, we recognize the power of storytelling transcends mere information transmission; it shapes beliefs, influences actions, and molds the collective conscience of societies. In this crucial chapter, we explore how the multifaceted stories of climate change are crafted and disseminated across media platforms, setting the stage for a culturally inclusive climate discourse that respects the nuanced perspectives of diverse communities. The capacity of these narratives to inspire action is undeniable.

We examine the art of merging scientific facts with emotive elements to galvanize communities, businesses, and policymakers toward a resilient and sustainable future. Moving beyond mere statistics and ecological models, the use of relatable anecdotes and impactful visuals not only broadens the reach of climate messaging, but also embeds the gravity of environmental stewardship into the fabric of daily life. Engaging diverse voices in climate discourse not only enriches the conversation, but also ensures marginalized narratives are no longer silenced but rather amplified to effect meaningful change. In harnessing the collective power of these stories, we can inspire a vision of hope and determination, shifting the zeitgeist from one of apathy to one of vibrant, actionable concern for our shared home.

Media Representation of Sustainability

In an era saturated with media at every turn—social platforms, 24-hour news cycles, streaming documentaries, etc.—the stories we encounter about climate change shape our perceptions and actions. The media's role in representing sustainability is pivotal. How these narratives are crafted can either galvanize society to take proactive steps or lull it into complacency or hopelessness.

At its most inspiring, media representation of sustainability presents a vision of a balanced world. It can showcase the beauty and resilience of the Earth, fostering a connection between viewers and their environment. This form of storytelling has the power to ignite a passion for conservation and a desire to live more sustainably.

However, the challenge for media is to present sustainability in a way that resonates across different cultures and societies. Media outlets must navigate complex scientific information and make it accessible without oversimplifying or sensationalizing. They walk a tightrope, balancing urgency with solutions and despair with hope.

The representation of sustainability issues can influence policy decisions and drive corporate behavior. When the media highlights the success of green businesses or the potential of renewable energy, it can shift the market and encourage investment in sustainable initiatives.

Positive media coverage of sustainability can also instigate behavioral changes. By showcasing practical measures individuals can take, such as reducing waste or conserving water, the media serves as a bridge between knowledge and action.

Documentaries have played a significant role in this sphere, bringing both the beauty and the destruction of the planet into people's living rooms. Yet, there is a balance to be struck. Too much emphasis on destruction can lead to "doom fatigue," where the scale of the problem paralyzes rather than motivates action.

Prevailing economic interests or political agendas can sometimes skew media reporting on sustainability. When ownership of media channels is concentrated, so too can be the viewpoints on sustainability and climate change. Thus, diversity in media leadership is critical for inclusive and balanced storytelling.

Media representation often mirrors society's current concerns about sustainability. Narratives of climate change have evolved from distant, future-oriented risks to immediate challenges affecting communities globally. Real stories of climate impacts have become a powerful tool in connecting the abstract concept of climate change to individual experiences.

One vital aspect of sustainability in the media is the inclusion of Indigenous and marginalized voices. Climate change affects these groups the most, yet their perspectives and solutions are frequently underrepresented. Including their stories not only brings equity to the conversation, but also presents a wealth of traditional knowledge in the practice of sustainability.

Further, visual media holds a considerable influence when it comes to telling the stories of climate change. The right image can become iconic, encapsulating complex ideas in a moment frozen in time. Think of the polar bear on the melting ice cap or a forest ablaze. These images provoke emotional responses and can become rallying points for collective action.

While the media can be a powerful force for change, the responsibility does not lie with them alone. Consumers also carry the mantle of critical engagement They are responsible for seeking out diverse sources, questioning narratives, and contributing to the discourse with their feedback and activism.

With social media's ascendancy, there's a democratization of sustainability narratives. Citizen journalists and activists can circumvent traditional gatekeepers and share their stories directly with global audiences. This trend allows for a more grassroots, bottom–up approach to shaping the sustainability discourse.

The interplay between media and education on sustainability is also of note. As educational institutions incorporate sustainability into curricula, media resources become vital teaching tools by offering vivid examples that can anchor theoretical knowledge in real-world contexts.

To maintain a fertile ground for sustainability narratives, media professionals must uphold good journalistic practices including critical investigation, balanced reporting, and a commitment to truth. As much as media reflects society, it also projects visions of the world that could be.

Conclusively, the media's role in sustainability is not merely descriptive but prescriptive. The representation of sustainability within media narratives is not just about reporting what is but inspires what could be. As storytellers of our time, media practitioners have the unique opportunity—and duty—to wield their craft in contributing to the Earth's renaissance.

Engaging Diverse Voices in Climate Discourse

The challenge of climate change is as varied and multifaceted as the human stories at its core. Its narrative doesn't belong to a single voice but rather is composed of a chorus of diverse perspectives, each holding a critical piece of the puzzle. In engaging with these varied voices, from the still whispers of isolated tribes to the loud deliberations in international assemblies, we stitch together a comprehensive tale. It's essential that this mosaic of narratives includes everyone, as it's only through inclusivity that the stories we tell about climate change become truly representative of the human experience.

For too long, a narrow set of voices—typically those from the most affluent sectors of the world—has dominated the discourse surrounding climate change. Yet, the impacts of climate change are not felt equally. Those with fewer resources and less political clout—often Indigenous peoples and marginalized communities—are disproportionately affected despite contributing the least to global emissions.

Recognizing their experiences is crucial in painting an accurate portrait of the crisis at hand. The stories we share about climate change shape policies, drive grassroots movements, and influence individual behaviors. When these stories exclude significant demographics, they limit empathy and understanding across audiences. Engaging diverse voices in climate discussion invites a richer, more nuanced understanding of climate impacts and solutions that may have otherwise remained unearthed. These narratives from varied cultural backdrops provide fresh viewpoints and innovative approaches to adaptability and mitigation strategies.

Indigenous voices, for example, carry with them traditional ecological knowledge that has been honed over millennia. This wisdom offers insights into sustainable land use, preserving biodiversity, and living in balance with nature. It's a body of knowledge that's been largely overlooked, but as we face unprecedented environmental challenges, it becomes more apparent how invaluable this Indigenous science is.

Similarly, women, who often bear the brunt of climate change due to gendered societal roles—especially in agricultural and water gathering—offer essential perspectives on resource management and community stability in the face of change. Their stories underscore the importance of gender equity in climate solutions and resilience-building.

Youth, too, are becoming increasingly vocal in the climate narrative. They bring urgency and a future-focused vision to the conversation, challenging the status quo and advocating for more aggressive action against climate threats. These younger voices are pivotal in maintaining momentum for change and bridging intergenerational knowledge and action for climate adaptation and mitigation.

Business leaders also have a pivotal role in climate discourse. As the engines of innovation and economic growth, they bring perspectives on sustainable industry practices, green technologies, and corporate responsibility. Their stories speak to the possibility of balancing profitability with planetary health, showcasing that economic development need not be at the expense of the environment.

The climate narrative is incomplete without the stories of policymakers, who draft the laws and regulations that can significantly curb emissions, protect natural resources, and drive sustainable development. Their discourse is one of balancing immediate political pressures with long-term environmental goals, navigating complex global negotiations, and making the tough decisions that move societies toward a low-carbon future.

To engage these diverse voices effectively, we must seek out platforms that amplify often-marginalized narratives. Media representation plays a critical role, as does the inclusion of diverse participants in climate conferences and decision-making bodies. Educational programs must expand their curricula to include climate stories from around the globe, enhancing the cultural competency of learners and preparing them to look at climate issues through a wide-angle lens.

Language and communication styles vary across cultures and communities, which means successful discourse requires a level of cultural fluency. Venues for dialogue should allow for different modes of expression, whether through spoken word, art, music, or storytelling. By facilitating a space where different forms of knowledge are respected and valued, we encourage engagement and collaboration among a broader audience.

Moreover, engaging diverse voices isn't just about who speaks, but also about who listens. It's an invitation for each of us to step outside our own experiences and actively seek out the stories of others. It's a commitment to listen deeply, with empathy and a willingness to learn. In this way, we become not just narrators of our own stories but active participants in a shared human story of survival and resilience in the face of climate change.

Social media and digital platforms offer unprecedented opportunities to connect and spread voices around the globe. Used strategically, they can dismantle barriers to participation, bringing the frontline experiences of climate change from the most remote corners of the planet directly into the global conversation. This democratization of the narrative is vital in cultivating a global understanding and response to the climate crisis.

However, the work doesn't stop at listening and connecting; it requires action. Stories inspire action, and it is through hearing and embracing the array of challenges, solutions, and experiences related to climate change we find the pathways toward truly sustainable practices. These diverse climate stories should serve as a call to action to rethink our individual and collective choices and to create a world resilient to the climatic challenges ahead.

In engaging diverse voices, we're reminded that the climate crisis, while universal, doesn't have a one-size-fits-all solution. Instead, it requires a tapestry of strategies woven together with strands of local knowledge, cultural insight, and individual experience. It is in this heterogeneous approach we'll find our strength and creativity to address the complex problem that is climate change.

The collective effort to shape the ever-evolving story of our planet is a testament to the power of diversity in overcoming the greatest challenges. When we make room for all voices, especially those typically unheard, we don't just enrich the narrative; we may find the keys to unlocking a sustainable future for all. Climate discourse that values diversity is, in essence, the best reflection of our shared humanity and shared Earth.

Chapter 16:
Technology and Tradition: A Delicate Dance

I n the preceding chapters, we've traced the rich pattern of cultural diversity and its nuanced contributions to sustainable practices. Now, as we navigate the tapestry of technology and tradition in this chapter, we delve into the intricate interplay that drives innovation while honoring ancestral wisdom.

Bridging the gap between age-old customs and cutting-edge progress requires a dance that is both delicate and daring, balancing reverence for the past with the exigencies of our present and the unknowns of our future. This melding of worlds is not merely a philosophical exercise but an essential strategy for achieving global sustainability.

Within these pages lies an exploration of how technological advancements can be inspired and informed by traditional knowledge, a relationship that enhances both the contemporaneity of our actions and the continuity of our cultural heritages. Moreover, a consideration for access and equity is fundamental as we choreograph a future where green technology is shared and shaped by all, not monopolized by a few. This chapter endeavors to demonstrate it's not about replacing what is old but instead about enhancing the synergy between the two realms and ensuring sustainable practices are as inclusive as they are innovative.

Innovations Inspired by Global Cultures

In the delicate interplay between technology and tradition, innovations inspired by global cultures emerge as powerful beacons guiding our path to sustainability. Across the vast tapestry of humanity's diverse practices and beliefs, there is a formidable pool of ingenuity that has long fostered harmonious coexistence with the natural world. This section delves into how contemporary innovation can not only learn from, but also build upon, these rich cultural legacies.

One such innovation is the renaissance of ancient water-harvesting techniques. In regions like India, traditional methods such as "Johads"—small earthen check dams—have been resurrected to combat water scarcity. By modernizing these age-old systems with contemporary engineering, communities are successfully reviving depleted aquifers and mitigating the impacts of droughts, proving historical practices can be mobilized to solve modern challenges.

The concept of "It Takes a Village" is finding its application in technology through collaborative approaches to innovation. Co-creation labs and maker spaces often draw on local knowledge and techniques to develop solutions that are finely tuned to specific environments. For instance, solar-powered equipment for agriculture in sub-Saharan Africa is being designed with farmers' insights, resulting in tools that are not only efficient, but also culturally accepted and widely adopted.

Cultural sustainability also ventures into the realm of architecture. Traditional home designs, from the earth houses of Africa to the bamboo constructions of Southeast Asia, are inspiring sustainable building practices globally. These designs offer not only a low carbon footprint due to the use of local materials, but also natural climate control, thereby reducing the need for artificial heating and cooling.

On the agricultural front, the practice of polyculture farming, prevalent among many Indigenous peoples, has informed the burgeoning field of permaculture. This nature-mimicking approach to cultivation results in resilient ecosystems that can sustainably produce food while enhancing biodiversity—a stark contrast to the monoculture systems that dominate industrial farming.

The world's ubiquitous mobile technologies offer an avenue for the encryption of traditional knowledge into digital form, which ensures its preservation and wide dissemination. In this way, mobile applications can deliver localized weather forecasts, agricultural advice, and market information in Indigenous languages, directly bolstering the resilience of communities while honoring their linguistic heritage.

Biomimicry is another domain where cultural respect for nature's genius converges with cutting-edge innovation. Termite mounds, for instance, have inspired the design of buildings that regulate temperature without external energy sources. By emulating the natural air circulation systems found within these mounds, architects are creating structures that stand as testaments to the blend of tradition and technology.

In the field of waste management, we see the return to principles of circularity that many cultures inherently practiced. The Zero Waste philosophy, gaining momentum in modern societies, echoes the waste-not ethic of Indigenous peoples who utilized every part of an animal or plant. By reframing "waste" as a resource, we're reengaging with ancient wisdom in which the end of one process signifies the beginning of another.

Making strides in healthcare, contemporary medicine increasingly recognizes the merits of traditional healing practices. Many widely used pharmaceuticals have their roots in plant-based remedies discovered by native peoples. By integrating traditional and modern approaches to health, we can create comprehensive care practices that are both scientifically sound and culturally sensitive.

Energy generation is another arena ripe with cross-cultural innovation. For instance, solar power technologies are incorporating designs informed by local aesthetics and social structures. In some Pacific islands, for example, solar panels are being integrated into the canopies of communal meeting spaces, ensuring they align with the cultural significance of those areas and promoting community buy-in.

Transportation technologies, too, are evolving with an eye toward cultural sustainability. Electric bikes and scooters are being tailored to fit the artistic flair and structural needs of different locales, making them a viable and accepted transport option that reduces reliance on fossil fuels and embraces local identity.

Cultural festivals around the world are becoming hotbeds for sustainable innovation, effectively transforming these celebrations into showcases of tradition fused with green technology. From solar-powered lighting at outdoor festivals to waste-reduction campaigns during mass gatherings, the confluence of celebration and sustainability is becoming increasingly prominent.

Faced with the growing need for sustainable textiles, innovations in fabric manufacturing are harking back to traditional methods of using organic materials and natural dyes. This revitalization not only reduces the environmental impact of the fashion industry, but also supports local economies and maintains cultural identity.

As we wind through the lanes of global ingenuity, the cultivation of sustainable landscapes rises as a blend of art and science. Urban farming initiatives are drawing on Indigenous crop rotation techniques to maximize yield and maintain soil health, all within the spatial constraints of city living.

Finally, in the realm of community planning and development, a sensitivity to the cultural context is imperative. By designing urban spaces that reflect the historical and social fabric of their locales, we foster a sense of belonging and stewardship among inhabitants. Initiatives like community-managed green spaces and culturally themed urban renewals stand as testaments to what's possible when technology pays homage to tradition.

Throughout this mosaic of innovations, it is evident that traditional knowledge is not to be archived as a relic of the past but is instead to be woven into the fabric of tomorrow's progress. As we continue to navigate the intersection of technology and tradition, let us remember embracing global cultures is not merely an act of preservation, but a catalyst for sustainable advancement.

Ensuring Access and Equity in Green Tech

The quest for sustainability increasingly pivots around the innovative applications of green technology. However, the fruits of these advancements must be accessible and equitable to ensure sustainability is not just the privilege of the affluent, but a universal right of all people. Access and equity in green technology are essential pieces of the puzzle as we perform the delicate dance between tradition and modernity.

Economic, geographical, and social factors often skew access to green technology. Wealthier nations and individuals are typically the early adopters of sustainable solutions, enjoying the benefits of cleaner air and energy security that such technologies can provide. To dance gracefully, steps must be taken to bridge this divide and ensure technologies—once exclusive—can be cascaded down to those less privileged.

Equity is equally critical in this equation. Green technologies must account for diverse needs and circumstances. Understanding this, designers and policymakers are tasked with developing solutions that are adaptable and sensitive to various cultural contexts. It's about providing tools for sustainability that mesh effectively with different ways of life without demanding a sacrifice of cultural identity.

Expanding the reach of green technology also entails subsidizing the costs for lower-income families and countries. There isn't a one-size-fits-all approach, but mechanisms such as sliding scale pricing models, grants, and low-interest funding solutions can make a tangible difference.

Educational initiatives are fundamental to leveling the technological playing field. To administer such technologies effectively, individuals and communities must be knowledgeable advocates and practitioners. Hence, education is a cornerstone of sustainability. Without it, even the most powerful tools can be rendered ineffective.

We confront the challenge of overcoming historical mistrust and resistance to new technologies in certain communities. To proceed thoughtfully, engaging community leaders and respecting traditional wisdom is vital. Successful integration of green technology must occur in tandem with conserving the integrity of local cultures and practices.

Rural areas often lag behind urban centers in adopting green technology due to issues such as inadequate infrastructure and information gaps. By providing mobile solutions, microgrids, and community-based projects tailored to rural realities, we can begin to patch these disparities.

There exists a moral obligation to design and manufacture green technologies that incorporate principles of circular economy, thus avoiding the creation of new forms of environmental degradation and exploitation in the pursuit of sustainability. It's not just what technologies are being used, but also how and why they are produced, that underscores the narrative of a just and durable future.

Considerations surrounding the life cycle of tech products also urge us to think critically about waste management. There has to be a strategic approach toward recycling and refurbishing to circumvent the generation of electronic waste, which is an increasingly alarming concern.

In the urban context, green technology can meet equity goals by regenerating neglected areas. Green roofs, living walls, and urban agriculture can transform impoverished urban landscapes into models of sustainable living, contributing to social cohesion and improved quality of life.

Engagement and collaboration with Indigenous peoples are critical to identifying how green technology can support, rather than undermine, their self-determination and sovereignty. Their knowledge systems, when honored and integrated, can provide valuable guidelines on how to adopt and adapt technologies in ways that are beneficial for the environment.

Assessment measures and reporting systems ought to be put in place to monitor whether green technology is being deployed accessibly and equitably. These evaluation protocols can provide accountability and transparency, guiding ongoing efforts in the right direction.

Partnerships across sectors are essential for sharing technology and expertise in order for knowledge to flow freely to where it's most needed. Public-private partnerships, for instance, can leverage resources, reduce risks, and increase the deployment rate of green technologies.

Lastly, we must address the economic structures that either enable or hinder the spread of green technology. This calls for a re-imagination of trade agreements, intellectual property laws, and subsidies that currently favor existing high-emission industries and technologies over their green counterparts.

Ensuring access and equity in green technology is not merely a deployment challenge but a profound moral and social paradigm shift. It's about ensuring the drive toward a sustainable world leaves no one behind. To achieve this, each step forward must be measured, inclusive, and reflective of the rich tapestry of global culture we aim to preserve.

Chapter 17:
Cultural Festivals of Sustainability

E mbarking on a celebratory journey, this chapter highlights how traditional and modern festivals across varied cultures have become a fervent ground for the promotion of sustainability and environmental consciousness.

Cultural festivals, with their dazzling display of heritage and unity, encapsulate an ethos that resonates with the sustainability movement. They're not just occasions for joy, but also for reflection, education, and action toward a more sustainable existence. These festivals harness the collective energy of communities to honor the Earth, contributing to a diverse and dynamic narrative of ecological stewardship. Rooted in age-old customs, yet vibrant with innovative ideas, these festivals exemplify living traditions that adapt and flourish without losing sight of their core ecological principles.

By engaging attendees in practices that emphasize conservation, such festivals have the potential to inspire sustainable lifestyle changes and create a ripple effect beyond their timelines, instilling lasting respect for our interdependence with nature. From the colorful regalia of Indigenous harvest festivals that embody the cyclical giving and receiving with the land to the urban eco-art exhibits fueling discussions on resource use and green living, these events act as microcosms where celebration and sustainability meld seamlessly, invigorating participants with a shared vision of maintaining our planet's balance for generations to come.

Celebrations That Honor the Earth

As we weave our way through the vibrant tableau of cultural festivals worldwide, we find profound examples where the celebration is synonymous with sustainability. Celebrations that honor the Earth are not merely festive occasions; they are bastions of cultural heritage that echo the profound relationship between humans and the natural world. These occasions are demonstrations of respect and a call to the collective consciousness, reminding us we are not conquerors of the Earth, but rather stewards tasked with nurturing its well-being.

Globally, many cultures mark the change of seasons with festivities that pay homage to the Earth's natural cycles. Agricultural societies, for example, have long had festivities aligned with sowing and harvest times. Such celebrations are steeped in gratitude; they recognize the bounty the Earth provides and the delicate balance required to maintain it. Through these rituals, communities pass on knowledge and respect for local ecosystems from generation to generation, thus maintaining a form of cultural sustainability that meshes seamlessly with environmental stewardship.

One such example is the Vernal Equinox, known in various traditions as Nowruz or Ostara. This celebration, which marks the arrival of spring, is a time for rejuvenation and the planting of new crops. Communities from the Iranian Plateau to Northern Europe observe practices such as cleaning homes to welcome new beginnings, planting trees, and preparing traditional foods that symbolize fertility and growth. These customs foster a connectedness to the land and a shared understanding that the well-being of our environment directly influences human survival and prosperity.

Another celebration deeply intertwined with environmental reverence is the festival of Pongal in South India. This four-day harvest festival thanks the sun, nature, and the farm animals that contribute to a bountiful harvest. Traditional practices include boiling the first rice of the season and offering it to the Sun God, symbolizing the interdependence between humans and nature in sustaining life.

In the Indigenous communities of the Americas, many ceremonies and festivals serve to honor the Earth. For instance, the Native American tradition of Pow Wow includes dances, music, and attire that carry deep ecological significance and often tell stories of animals, water, and the Earth. The Pow Wow promotes unity and cultural exchange, emphasizing respect for nature as central to the community's identity and existence.

Similarly, in many African cultures, festivals act as a conduit for educating the young about traditional environmental knowledge. The rituals and stories shared during these times bind people to their environment, signifying a reciprocal relationship where the protection and celebration of biodiversity are imperative for cultural preservation.

In Japan, the practice of Hanami, or flower viewing, particularly of the cherry blossom, is a centuries-old tradition that commences with the blooming of sakura trees. It is a celebration of natural beauty and the ephemeral nature of life, reminding us of the fleeting and precious nature of our resources. Hanami inspires a collective cherishing of the moment, fostering appreciation and—by extension—the conservation of these natural wonders.

The role these celebrations play in sustainable living cannot be overstated. They bolster community ties and reinforce the notion that we share a finite planet. The emphasis on harmony with nature underscores the importance of sustainable practices in everyday life, from agricultural methods to water use to waste management. Festivals thus transition from mere temporal events to year-round mindsets where sustainability is interlaced with cultural identity and practice.

For businesses and policymakers, understanding and supporting these cultural festivals can be critical in creating effective sustainability strategies. Recognition and amplification of these festivals invite a broader audience to engage with sustainability in a cultural context. They provide a platform for education and inspiration, making the abstract concept of "sustainability" tangible through tradition and celebration.

These festivals also highlight the importance of biodiversity, often featuring native plants, animals, and traditional agricultural practices tailored to local ecosystems. This diversity is essential for resilience in the face of environmental challenges. Conserving the variegated threads of life is not only an ecological imperative, but also a cultural one.

Yet, as we celebrate, we must also confront the reality that many of these traditions are under threat from global environmental changes. Climate change, habitat destruction, and cultural homogenization pose significant risks. In this context, festivals become a platform for resistance and resilience, assertively reminding us of the need to safeguard the planet for future generations.

Take, for example, the celebration of the Lunar New Year across various Asian cultures, which is increasingly incorporating sustainable practices such as minimizing fireworks to reduce air pollution and opting for electronic red envelopes to lessen waste.

As these festivals evolve, we find they continue to foster the symbiotic relationship between culture and sustainability. Within the joyous cacophony of music, dance, and color, there lies a deeper narrative—a manifesto of conservation and respect for the Earth. They beckon us to join in the celebration not as bystanders, but as active participants in the ongoing stewardship of our planet.

In summation, these celebrations are more than dates on our calendars; they are invitations to learn from the past, engage with the present, and act for the future. They are a canvas depicting the story of human interdependence with nature—each festival a brushstroke, each tradition a color, contributing to a masterpiece of sustainable living. To honor the Earth is to engage in ritual acts of continuity, ensuring the rhythms of nature will go on inspiring and sustaining humanity as they have done since time immemorial.

Learning from Festive Practices

Cultural festivals hold a mirror to society's values and practices, reflecting deep traditions that can inspire sustainability. Within the beating heart of these festivities lie invaluable lessons on community resilience, environmental stewardship, and the symbiotic relationship between humans and nature. In this section, we delve into how festive practices contribute to sustainability and what we can learn from them to reinforce our commitment to a thriving planet.

The concept of sustainability is not new. Many cultural festivals date back centuries and are inherently sustainable, designed to cycle with the natural seasons and rhythms of the Earth. They teach us the art of balance—how to take and how to give back in order to ensure nature's bounty remains for future generations. This wisdom, practiced and preserved, must be understood and applied to current sustainability efforts.

Festivals often emphasize themes of renewal and regeneration. Agriculturally based celebrations, for instance, honor the harvest and the sowing of seeds—both literal and metaphorical. They encourage communities to consider the source of their sustenance and the means through which it proliferates. This acknowledgment of natural cycles fosters respect for the elements that sustain life and serves as a reminder of our dependence on these systems.

Moreover, many festive practices involve communal sharing and minimizing waste, central tenets of modern sustainability. Potlatches of the Indigenous peoples of the Pacific Northwest involve giving away possessions and sharing resources, thus fostering a culture of generosity and circular economy principles that predate their mainstream acceptance by millennia.

Sustainability also extends to the preservation of culture and knowledge. Celebrations act as vessels for intergenerational knowledge transfer, with elders imparting time-honored methods and beliefs to the youth through stories, songs, and rituals. These interactions are key to sustaining cultures and ensuring ecological practices do not fade away but instead evolve with time.

At the local level, festivals can reinforce community cohesion, demonstrating how collective efforts lead to greater outcomes—whether it's in the organization of events or the conservation of local habitats. For instance, the community-driven nature of festivals instills a sense of ownership and responsibility toward the locality, which is crucial in maintaining sustainable practices.

The creativity and innovation displayed during festivals represent a resourcefulness that can be applied to sustainable developments. The way materials are reused and recycled for decorations and costumes offers lessons in waste reduction and the imaginative revalorization of items that would otherwise be considered trash.

On a broader scale, cultural festivals tend to involve eco-friendly practices, such as the use of local ingredients for food and materials for crafts, minimizing transportation emissions and promoting local biodiversity. These practices prioritize local ecosystems and economies, reinforcing the need for self-sustainability.

Moreover, festivals often inspire collective action toward large-scale environmental initiatives. As public gatherings that can command attention, they provide a platform for awareness-raising and environmental advocacy, galvanizing community support for conservation efforts and policy changes.

As we continue to face the challenges of climate change, festivals can be harnessed as tools for instilling a sense of urgency and motivation. Just as a festival can excite and mobilize a population, so too can the principles behind these practices invigorate our approach to sustainable living, pushing us beyond apathy and into action.

Taking cues from the traditional practices embedded within festivals around the world provides concrete examples of how sustainability can be naturally integrated into the fabric of daily life. They demonstrate the feasibility of living in harmony with our environment while celebrating our cultural heritage.

Festivals, at their core, bring communities together, and it is this unity that may prove to be our greatest strength in the quest for sustainability. The shared experiences and communal sense of accomplishment in organizing and running these events kindle a spirit of collaboration that is essential to sustainable progress.

As we draw inspiration from these festive practices, it's crucial to remember sustainability is not a one-size-fits-all solution. What works for one culture or community may not be suitable for another. However, the underlying principles of resourcefulness, balance, and respect for nature are universal values that can guide us collectively toward a more sustainable future.

By embracing festive practices as teaching tools, sustainability becomes not only an act of preservation but a celebration. A shift in perspective from restriction and sacrifice to joy and reverence makes the path to sustainability one of positivity and hope, mirroring the celebratory spirit that festivals represent.

In conclusion, cultural festivals—whether they are marked by the vibrant hues of Holi, the reflective tradition of Thanksgiving, or the endurance of the First Nations people of Australia corroboree—are more than just occasions for joy. They are repositories of practices and philosophies that can guide our approach to sustainable living. By approaching these festivals with an eye for learning, we ensure we do not merely observe traditions; we evolve with them, carrying forward their most enduring and valuable lessons into our global efforts to sustain our planet's future.

Chapter 18:
Culinary Cultures: A Taste of Sustainability

I n "Culinary Cultures: A Taste of Sustainability," we delve into the rich tapestry of food traditions that not only tantalize the palate, but also embody sustainable practices shaped by centuries of cultural wisdom. The ways in which communities around the globe harvest, prepare, and share food are deeply rooted in a complex interplay of environmental stewardship, social cohesion, and economic viability. Food is more than sustenance; it is a medium through which cultural identity is expressed and preserved.

By examining diverse gastronomic practices, this chapter explores how traditional diets can contribute to a sustainable future, highlighting the potential for locally sourced, seasonally-aligned eating habits to reduce our ecological footprint while enhancing nutritional health. Through the lens of sustainability, we learn how the principles of diversity, moderation, and mindfulness at the dining table can drive positive change in our food systems. By embracing the symbiotic relationship between human well-being and planetary health, we can discover how culinary cultures offer invaluable insights into cultivating a more sustainable and resilient world.

Food Traditions Shaping Sustainable Diets

In the narrative of sustainability, the cultural significance of food traditions plays a vital role. Commensal rituals and age-old recipes serve not only as a medium for sustenance, but as an edifice of identity and community. These traditions, steeped in the wisdom of generations, can point us toward a future where our diets are as sustainable as they are nourishing.

Around the globe, food practices bound to cultural identities have shown a resilience from which modern agricultural methods can learn. The reliance on seasonal produce, local sourcing, and biodiversity found in traditional diets demonstrates a low-impact approach inherently aligned with environmental conservation. Communities that maintain their traditional foodways often possess a deep understanding of the land and its capacities, which helps to create a sustainable balance between humans and nature.

Anchored in the ethos of respect and reverence for the land, Indigenous cultures have long understood the interconnectedness of all life forms. The intricate knowledge of agroecology—how plants and animals interact within agricultural systems—is a treasure trove for advancing sustainable diets. By embracing these holistic perspectives, we can foster agricultural practices that are both productive and regenerative.

The act of growing and preparing food according to ancestral knowledge is also a statement of resistance—resistance against a homogenized food system that often prioritizes convenience over quality and diversity. By keeping alive the variety of crops and animal breeds that are well-adapted to local environments, we safeguard genetic diversity essential to resilience against diseases and climate change.

Take, for instance, the Mediterranean diet, characterized by a high consumption of vegetables, fruits, legumes, and olive oil and a low consumption of meat. It's not only recognized for its health benefits, but also for its lower environmental footprint when compared to Western dietary patterns. The embedded emphasis on plant-based foods and minimal meat consumption echoes sustainable dietary guidelines that promote reduced greenhouse gas emissions.

The role of the community in food traditions is imperative. Festivals, agricultural fairs, and farmers' markets are sociocultural events that foster community ties and celebrate local cuisine. These gatherings often encourage the consumption of local, seasonal produce, reinforcing the link between community culture and a sustainable food system.

Many traditional food systems are inherently waste-conscious. The nose-to-tail approach in meat consumption, where no part of the animal is wasted, and the use of every part of a harvest exemplify efficient utilization of resources. These practices, entrenched in cultural food traditions, highlight how waste minimization can be a natural part of our dietary habits.

However, sustaining these dietary traditions in the modern world requires integration and adaptation. Globalization and urbanization present both challenges and opportunities. While the movement of people and ideas can lead to the erosion of traditional foodways, it can also spread knowledge of sustainable practices.

Younger generations play an essential role in the transmission of food heritage. By linking educational initiatives with culinary traditions, we can impart the importance of sustainable diets to upcoming cohorts. Initiatives such as school gardens and cooking classes grounded in traditional food knowledge can inspire young people to carry forward their culinary heritage while being mindful of the environment.

Educating consumers about the impacts of their food choices on the environment and on their health is key. By raising awareness of the environmental cost of processed and imported foods compared to those grown locally and prepared traditionally, we can shift attitudes and behaviors toward more sustainable consumption patterns.

Policy frameworks can support and promote sustainable food traditions by recognizing and protecting traditional agricultural practices, supporting small-scale farmers, and facilitating access to markets for traditional and local products. Such measures can ensure sustainable food practices are not only preserved, but also encouraged.

Businesses also have a role in upholding sustainable diets through ethical sourcing, transparent supply chains, and supporting local farmers and artisanal producers. By aligning business practices with cultural sustainability, they not only contribute to a more resilient food system, but also enrich the cultural fabric of our diets.

Finally, chefs and food influencers who spotlight traditional ingredients and preparation methods can continue to shape public opinion and culinary preferences. Their platforms can be powerful tools for reinvigorating interest in heritage foods and sustainable eating practices.

In sum, the path to a sustainable future is paved with the wisdom of our culinary past. By supporting food traditions that are in harmony with the Earth, nurturing the link between culture and the table, and embracing the power of community, each meal can be a step toward a more sustainable and vibrant world.

The Role of Gastronomy in Conservation

In the wake of a burgeoning global awareness, the integral role of gastronomy in the conservation of biodiversity and ecosystems is garnering increased attention. Gastronomy, the art and knowledge of food and cooking, is a potent tool in conserving the diversity found within our landscapes and on our plates. It's through culinary innovation that preservation is possible, transcending the borders of science and culture and engaging multidisciplinary approaches for sustainable development.

At the heart of gastronomy is the provenance of ingredients, which speaks to the geography, climate, and culture of a place. Chefs and food enthusiasts who champion local ingredients are inadvertently supporting the conservation of local ecosystems and agrobiodiversity. Their choices, which influence the broader populace, help safeguard native species and traditional farming practices that are under threat from industrial agriculture and homogenized food systems.

Traditional food systems are repositories of a vast array of genetic diversity. From heirloom vegetables to Indigenous breeds of livestock, each has been adapted over centuries to thrive in specific climatic conditions and ecological niches. Gastronomy encourages the use and maintenance of this diversity by celebrating Indigenous ingredients within culinary traditions, providing sustainable livelihoods and conserving vital resources.

Renowned chefs have the power to influence consumption patterns by directly impacting food trends. When these culinary leaders commit to sourcing sustainably produced, local, and seasonal ingredients, they drive consumer demand in that direction. This direct action promotes agricultural practices that are in harmony with local ecosystems and advances sustainable farming strategies.

Moreover, gastronomy has the potential to rekindle interest in "forgotten foods"—varieties that were once commonly cultivated but have fallen out of favor in the modern palate. The re-introduction of these foods can boost agrobiodiversity and provide nutritional variety, all while offering new economic opportunities for smallholder farmers.

The gastronomic tourism industry, which encourages travelers to explore regions through their unique culinary heritage, also plays a notable role in conservation. By creating a market for traditional dishes and local specialties, gastronomy stimulates the demand for biodiversity on a plate. This ecological gastronomy turns food into an ambassador for conservation, celebrating local ecosystems and cultural heritage.

New movements like "slow food" are marrying the pleasures of the table with a commitment to community and the environment. They emphasize the importance of preserving local culinary customs and ingredients, understanding that the slow, careful enjoyment of food is intrinsically linked to the health of the natural world.

Educational initiatives that combine gastronomy with environmental science can instill a deep appreciation for the interconnectedness of our food systems with natural cycles. Courses that teach the fundamentals of sustainable agriculture, along with the culinary arts, empower new generations to become champions of both delicious cuisine and the Earth that provides it.

Communities around the world have long understood the harmony that has to exist between their food practices and the lands they inhabit. Gastronomy's role in conservation is as much about preserving these knowledge systems as it is about protecting physical species and environments. It's a realm where taste and sustainability are not at odds but are threads of the same narrative.

Policy frameworks that recognize and incentivize the conservation efforts of the gastronomic sector can amplify its positive impact. Financial support for the use of Indigenous crops and sustainable hunting practices can translate gastronomic endeavors into a broader movement toward ecological stewardship.

Gastronomy, with its hub of flavors, traditions, and innovative practices, serves as a bridge between the past and the future. It helps to maintain cultural identities while fostering a relationship with nature that is symbiotic and endurant. Aptly, it's through the shared language of food that conservation efforts can resonate most profoundly with the broadest possible audience.

In summary, as we navigate the challenges of modernity and seek sustainable paths forward, the role of gastronomy in conservation cannot be understated. Its ability to connect culture, ecology, economy, and community makes it an essential ally in the endeavor to craft a more resilient and flavorful world.

Chapter 19:
Fashion and Fabrics: A Material Conscience

As the fabric of society becomes increasingly aware of its environmental footprint, the narrative of fashion and textiles is being meticulously rewoven to embody a material conscience. The very clothes that drape our bodies have deep cultural and environmental implications, which we often take for granted. Every fiber of their being is interlaced with the potential to either harm or heal our planet.

This chapter delves into how a rekindled respect for traditional practices, coupled with cutting-edge innovations in fabric production and waste reduction, is fostering the evolution of an industry that aligns aesthetic appeal with ethical accountability. It explores the growing tapestry of eco-friendly apparel initiatives worldwide and the invaluable role they play in cultural expression while lessening their impact on the Earth.

For those who adorn themselves with the latest trends, this chapter isn't just about making a fashion statement; it's a call to action, urging readers to don the mantle of change and to choose garments that not only look good but do good. By embracing sustainable fashion, each of us holds the power to influence industry giants and define our culture through the conscious selection of what we wear and igniting a broader transformation toward a sustainable future.

Eco-Friendly Apparel from Around the World

As we have unraveled the intricate patterns of cultural contributions to sustainability in previous chapters, we now switch our focus to the fabric of global fashion. The world of eco-friendly apparel extends far beyond our local boutiques and online shops; it is a resplendent tapestry that weaves together diverse threads of environmental consciousness from around the globe. Moving toward sustainable fashion is not just an industry trend, it's a heartfelt response to the environmental cries of our planet. It's about reimagining the life cycle of clothing from design to disposal, ensuring every garment we drape over our skin is a testament to our respect for the Earth.

Delve into the heart of India, where traditional handloom textiles embody centuries-old craftsmanship while embracing eco-friendly practices. This is not merely production; it's a ritual of patience and precision, respecting the loom and the natural fibers that pass through it. Here, dyes such as indigo, turmeric, and henna, are derived from the Earth, leaving imprints that tell stories of the land. The Indian fashion industry is increasingly acknowledging the value of these practices, making them an integral part of their sustainable fashion lines.

In Scandinavia, minimalist fashion meets environmental responsibility. Swedish and Danish designers are at the vanguard, creating apparel with the ethos of "less is more." Their designs reflect a deep understanding of longevity and waste reduction, prompting consumers to invest in pieces that endure both style and time. Using organic materials, legislating for textile recycling, and advocating for a circular fashion economy are just the beginning steps for this region's commitment to green clothing.

Across the vast expanse of Africa, artisans are spinning sustainability into every fiber of their creations. In countries like Ghana and Kenya, repurposed materials become vibrant fashion statements, expressing both cultural heritage and ecological foresight. These communities are not simply "following" a trend; they are leading the movement by designing apparel that holds the Earth as sacred and utilizing local and recycled resources to craft garments that shimmer with life-affirming energy.

Moving to South America, the Andean tradition of alpaca wool provides warmth without the environmental toll often associated with the textiles industry. Alpaca herding and fiber production are low-impact practices rooted in a harmonious relationship with the ecosystems of the Andes Mountains. Designers in Peru, Bolivia, and Ecuador are marrying ancestral techniques with modern sustainability standards in order to produce luxurious textiles that support local livelihoods and protect delicate biomes.

In the East, Japanese innovation in sustainable textiles is reshaping the way we perceive "eco-friendly." By creating new fabrics from repurposed and regenerated materials and embracing a culture of meticulousness and conservation, Japanese designers are crafting clothing that speaks volumes of their reverence for efficiency and environmental harmony.

Nestled within the innovation of eco-friendly apparel are initiatives like zero-waste pattern-making, a technique that eliminates textile waste during the design process. Pioneering designers in places such as New Zealand and Canada are proving a garment's life can be circular—where the end of one piece can be the beginning of another. This approach challenges the disposable culture endemic in the fashion industry, begging each of us to consider the longevity of our wardrobe choices.

Realizing the power of collaboration, global fashion collectives are forming—with designers, farmers, and artisans coming together from different corners of the world. These collaborations pave the way for a collective vision of eco-friendly fashion, pooling wisdom, creativity, and resources to produce apparel that boasts international sustainable standards.

Let's not forget the rise of upcycling movements. From the bustling streets of New York to the colorful markets of Bangkok, creative souls are transforming pre-loved items into unique, eco-conscious apparel. This practice is not only eco-friendly, but also a powerful statement of individuality in an industry often criticized for its homogenizing effect.

However, the journey toward sustainable fashion does not rest solely on the shoulders of designers and manufacturers. As consumers, we play a pivotal role in demanding transparency and sustainability in our clothing, and it is through our purchasing power that we encourage the industry to favor green practices. By choosing eco-friendly apparel, we stitch our own patch into the quilt of global sustainability.

To catalyze this transformation, we see innovative online platforms promoting eco-friendly fashion, connecting ethical producers to a global audience yearning for conscious clothing. These digital marketplaces are not just sales venues; they are educational hubs where stories of sustainability are shared and the virtues of conscious consumerism are affirmed.

Eco-friendly apparel is more than a commodity—it is an expression of cultural respect, environmental stewardship, and a commitment to ethical practices that uplift communities and preserve our shared habitats. As we adorn ourselves with the world's myriad of sustainable fabrics, let us remember each garment carries with it a narrative of its origin, a story that pays homage to the resources and hands that crafted it.

We must also consider the impacts of our clothing after its life on our backs. A sustainable world necessitates responsible disposal of apparel—be it through recycling, composting, or repurposing. Progressive cities and countries are implementing programs and policies to foster responsible clothing life cycles, inspiring us to be mindful of our wardrobe's environmental footprint even after its use.

As we clothe ourselves for the future, let us envision a world where every thread, button, and zipper embodies the sustainable spirit of our planet—a world where fashion is a tapestry of eco-friendly threads woven from the heart of every culture, telling the enduring story of a resilient Earth. Our choices in fashion are statements of our values, and by choosing eco-friendly apparel, we vouch for a world that treasures and protects its precious weave of life.

Cultural Expression Through Sustainable Fashion

Sustainable fashion represents an intersection where culture, creativity, and environmental consciousness converge. The fabric of our societies is woven with threads of identity and expression, and nowhere is this more evident than in the clothes we wear. Through sustainable fashion, individuals and communities around the world articulate their cultural heritage while also making a definitive stance toward preserving the Earth. This section unravels the rich tapestry of cultural expression interlaced with the fibers of eco-conscious apparel.

In embracing sustainable fashion, we are not simply choosing environmentally friendly materials; we are participating in a storied tradition of valuing resources and honoring the craftsmanship that draws from the diverse pools of cultural knowledge. Each garment, shaped by sustainable practices, carries a narrative of heritage reflecting the world's multitude of identities. By upholding these practices, we knit together past wisdom with modern exigencies, crafting an attire that speaks to the soul as much as to societal needs.

Indigenous communities, for example, have long demonstrated a profound symbiosis with their natural environments, and this ecological awareness is apparent in traditional attire that utilizes organic dyes, local fibers, and culturally specific designs. By incorporating these elements into contemporary fashion, we not only preserve this knowledge, but also propel it into new realms of relevance. It's a celebration of diversity that also serves the pressing demands of sustainability.

Historically culpable for considerable environmental degradation, the fashion industry is now being reimagined through the lens of cultural preservation and environmental responsibility. Sustainable fashion advocates for minimal waste, responsible sourcing, and ethical labor practices while allowing cultural identities to flourish. It's a radical departure from fast fashion, a model rooted in disposability and profit, toward one that sees the longevity of culture and the planet as intertwined.

Materials play a critical role in this sustainable fashion movement. Organic cotton, hemp, bamboo, and recycled fibers are becoming the standard bearers of this change. Each material has a cultural and ecological backstory, whether it be the ancient use of hemp in Asian textiles or the contemporary revitalization of alpaca wool in Andean communities. These materials merge ecological sensibility with cultural homage, crafting garments that tell the broader story of a region's heritage and its commitment to the Earth's welfare.

Cultural expression thrives when it serves a greater purpose beyond aestheticism. Sustainable fashion becomes a platform for raising awareness, prompting conversations, and reflecting collective values. Designers and wearers alike are carving spaces for traditional patterns, motifs, and colors that communicate specific cultural narratives while advocating for sustainability. This alignment of ethical production with cultural specificity serves as a bold statement of identity and responsibility alike.

The adoption of sustainable fashion is not without its challenges. The high costs of sustainable materials and the labor-intensive processes of traditional techniques present economic obstacles that must be managed. Yet, these same challenges also inspire innovation. Artisans and designers are collaborating to find ways of scaling sustainable practices without compromising cultural integrity or environmental ideals.

Global fashion events have begun showcasing sustainable collections, with an increasing number of designers integrating ethical practices into their work. These displays offer a glimpse into a future where style and sustainability are partners in dance. They illustrate how cultural expression can be both timeless and timely, capable of honoring heritage while also safeguarding the planet for future generations.

Fashion education is also pivoting to incorporate sustainability into its core teachings. As the next generation of designers is steeped in an ethos of responsible creation, the integration of cultural expression within these paradigms is essential. Students are learning that understanding the cultural contexts of materials and designs is key to creating fashion that resonates on a deeper level.

Consumers, too, play an integral role in the shift toward sustainable fashion. By choosing culturally resonant, sustainably made pieces, they drive demand for clothing that reflects both their personal narratives and their environmental values. Through these purchasing choices, they become patrons of an art form that dresses the body as well as the spirit.

Grassroots movements have further fueled the sustainable fashion trend, with communities and cooperatives forming around the shared goals of cultural preservation and environmental protection. These collectives work toward developing sustainable fashion as an engine for local economic growth, creating jobs rooted in traditional skills while adhering to environmental best practices.

As we move deeper into the Anthropocene, the story of sustainable fashion is one that must feature prominently. It is a narrative of resilience, an ode to the intricacy of culture, and a testament to the ingenuity required to align our aesthetic desires with the planetary boundaries we face. Sustainable fashion is not just a material choice, it is a cultural statement and an active participation in the creation of a more resilient and beautiful world.

By entwining the strands of cultural expression with those of sustainability, we fashion a new ethos, one in which our collective tapestry is both diverse and resilient. This multithreaded approach can stitch together a narrative of hope and action that is as varied as the cultures it represents. It's a vision of fashion where the beauty we adore is matched by the beauty of our actions toward the world.

In conclusion, the dialogue of cultural expression through sustainable fashion is one that weaves together aesthetics, ethics, and tradition. In this tapestry, every thread counts, every pattern tells a story, and every color is a symbol of our shared commitment to the Earth. As we each fashion our future, let our garments reflect the world we hope to see—a world vibrant with cultural diversity and rich in environmental integrity.

Chapter 20:
Activism and Advocacy: Voices for the Earth

n the currents of change, where the actions of one ripple around the globe, activism, and advocacy arise as indispensable forces in the tidal push toward sustainability. "Activism and Advocacy: Voices for the Earth" unfolds the narrative of those bold and impassioned individuals and collectives who have raised their voices not merely to speak, but to be heard and to effect real change. These environmental champions come from varied backgrounds, yet they share a common thread: the unyielding determination to stand for the Earth.

This chapter delves into how grassroots movements ignite sparks that can light the fires of global solidarity, acknowledging that it's only through the collective whispers and roars of advocacy that the message of sustainability can reverberate with profound impact. Storytelling emerges as a potent tool for activism, transforming individual experiences into powerful catalysts for broader societal engagement. The intricate dance between local actions and global repercussions is explored, emphasizing the importance of interconnected standpoints in fortifying the resilience of our planet. This examination of the dynamics of activism serves not only to inspire readers, but also to equip them with the knowledge that their voice, too, is a vital instrument in this harmonious chorus advocating for the Earth.

Grassroots Movements and Global Solidarity

As the world's ecological banner unfurls, a chorus of grassroots movements echoes across continents, stitching together a tapestry of environmental advocacy that testifies to the strength of collective action.

162

The term "grassroots" illustrates the organic, community-level origin of these efforts, where individuals and local organizations seed the very change they wish to see in the world. This section explores the burgeoning kaleidoscope of grassroots movements, where diverse voices coalesce to form a symphony of global solidarity for the Earth.

At the heart of these movements is the recognition that our environmental challenges are not bound by borders but are instead shared burdens that require communal resolve. As campaigns flourish, they often transcend their local confines, galvanizing allies on an international scale. The power of grassroots movements lies in their ability to mobilize ordinary people, transforming bystanders into activists with a shared conviction that every voice can shape the future of our planet.

Solidarity, a term that resonates with unity and mutual support, becomes the guiding force as local campaigns gain global resonance. Transnational environmental networks emerge, linking activists around the globe who, while rooted in diverse cultural soils, share a common vision of sustainability and justice. Through shared struggles and collective victories, these alliances illustrate the potential of interconnected efforts to tackle global crises.

One pertinent example is the proliferation of climate marches, where millions worldwide have taken to the streets to make their footsteps and placards a shared language of urgency. These mass mobilizations are not just demonstrations of public sentiment but catalysts for policy change through pressuring governments and corporations to heed the environmental clarion call.

Local grassroots movements often begin with an intimate understanding of their own ecosystems and a profound respect for the land that sustains them. This reverence is interwoven with pragmatic approaches to conservation and restoration, providing a beacon that lights the way for larger, more systemic change. Their actions speak volumes, embodying the sentiment that protecting our environment is not merely an act of charity but one of survival.

Global solidarity is also evident in the proliferation of information and resource-sharing. As grassroots movements navigate their unique challenges, they draw strength from a wellspring of shared knowledge, turning to one another for guidance, support, and inspiration. This exchange of strategies and successes forges a web of empowerment, buttressing local efforts with global wisdom.

Digital platforms have amplified this capacity for global solidarity, allowing activists to communicate instantaneously and cast a spotlight on issues that may have once lingered in the shadows. Social media campaigns can ignite an international dialogue, cohering fragmented narratives into a unified call for action. This digital dimension has expanded the battlefield for environmental advocacy, allowing for the mobilization of online armies in the defense of our natural world.

While grassroots movements are diverse in their tactics and objectives, many embody a commitment to direct action—a willingness to intervene at the sites where environmental degradation is most tangible. From tree-sits to blockades, from community gardening to water protection, these hands-on efforts resonate globally, setting a standard for environmental stewardship that transcends cultural and geographic boundaries.

Environmental justice movements blend grassroots vigor with global solidarity, acknowledging that the impacts of pollution and climate change are disproportionately borne by marginalized communities. These movements seek not just ecological redemption but social transformation, advocating for policies and practices that uplift both people and the planet.

Within this global patchwork, Indigenous activists play an instrumental role in advocating for sovereignty over their lands and the protection of age-old ecosystems. Their struggles often become emblematic of broader environmental conflicts, drawing support from a worldwide audience attuned to the intrinsic value of traditional knowledge and sustainable practices.

The solidarity witnessed on a global scale is a testament to the power of empathy and shared concern for our planetary home. As grassroots movements emerge from the local loam, they remind us we are all caretakers of the Earth, each of us responsible for weaving our strand into a resilient environmental fabric.

Yet, challenges remain. Ensuring cohesion while respecting the autonomy of distinct movements requires delicate coordination. There's a fine balance between maintaining a unified front against environmental threats and tilting toward a one-size-fits-all approach that could stifle the rich diversity of local campaigns.

Progress is measured not just in the victories won but in the relationships built along the way. Solidarity across cultures and continents fosters a sense of global citizenship, where each act of environmental advocacy—regardless of its origin—contributes to a collective narrative of hope and determination.

In embracing global solidarity, grassroots movements also face the challenge of sustaining momentum over time. Environmental victories are often hard-won and even harder to maintain. Here, the global community offers not just a cheering crowd but a continuing source of reinforcement, ensuring isolation or fatigue never extinguished local activism fires.

The unyielding spirit of grassroots movements, fortified by global solidarity, serves as a beacon of what can be achieved when the many sing as one for the Earth. As walls of apathy are dismantled and bridges of cooperation are built, we find that every act of environmental stewardship and every effort toward sustainability, is a stone laid on the path toward a resilient, thriving planet. It's a journey we embark on together as defenders of our shared home, united in purpose and hope for the world we wish to pass on to future generations.

Storytelling as Activism

In the realm of environmental advocacy, narratives have a unique power to inspire empathy, galvanize communities, and bridge the gap between knowledge and action. Storytelling as activism has emerged as a vital tool for those at the frontlines of ecological defenses worldwide. It's a practice that captures not only the imagination, but also the will of the people, mobilizing them toward a common purpose of protecting our planet.

Through the vivid and compelling power of stories, activists can shine a light on the often-ignored voices, particularly those of the Indigenous populations whose lives are intricately connected to the lands and waters under threat. One of the most compelling of these voices is that of Nemonte Nenquimo, a Waorani leader from the Ecuadorian Amazon.

Nenquimo's efforts exemplify how Indigenous storytelling can be an act of resistance and a rallying cry for environmental protection. As a co-founder of Amazon Frontlines and the Ceibo Alliance, Nenquimo has been at the forefront of a historic legal battle to save her people's ancestral territory from oil extraction. Her approach intertwines traditional knowledge with contemporary advocacy, engaging audiences worldwide through a narrative of survival and resilience.

A particularly potent aspect of Nenquimo's activism is its rootedness in personal experience. By recounting the Waorani's deep relationship with their land and the catastrophic consequences of oil pollution on their way of life, she doesn't just inform; she invites listeners to witness the urgency and injustice through a vivid emotional lens.

This tactic of using storytelling to connect personal battles to a broader struggle for environmental justice can be seen in various forms worldwide. From the tales of pastoralists in sub-Saharan Africa to the fishermen in the South Pacific, stories help to build a tapestry of activism that transcends geographical and cultural boundaries.

Indigenous stories often illuminate the interconnectedness of ecosystems in ways that transcend scientific discourse. This holistic perspective is essential to foster a broader understanding of sustainability, ecosystem health, and the complex impacts of industrial development. Nenquimo's storytelling, for instance, doesn't just highlight the threat to the rainforest, but also the potential loss of biodiversity and cultural heritage and the global impact of diminishing natural carbon sinks.

In the past, these stories might have been relegated to the periphery of environmental discourse. However, with the rise of digital communication platforms and social media, the voices of Indigenous activists are reaching a larger audience than ever before. Their narratives have become deeply influential in shaping public opinion and policy.

The power of such narratives also lies in their ability to foster identification and empathy in listeners. When people can see the human face of climate change and environmental degradation—the farmers whose crops are failing, the children whose health is compromised by pollution—they are more likely to understand the immediacy and severity of these issues.

Nenquimo's advocacy serves as a reminder that while statistics about deforestation and climate change are critical, they can sometimes be too abstract to provoke action. Personal stories transform these statistics into tangible, human experiences that people can relate to and rally around.

But storytelling as activism isn't merely about galvanizing public support—it's also an act of preserving heritage. As Nenquimo shares her people's ancestral wisdom, she is also ensuring their knowledge systems, cultural practices, and language are not forgotten amidst globalization and environmental crises.

These stories, however, are not just tales of loss and warning. They are also narratives of hope and problem-solving. Activists worldwide are using storytelling to share successful strategies for living sustainably, such as traditional water conservation techniques or agroforestry practices, providing invaluable insights for ecological management.

Storytelling as activism is also about creating a shared vision of a sustainable future—one that incorporates equity, respect for all life forms, and the preservation of cultural and biological diversity. Nenquimo's vision, for example, is not just a story of protecting the Waorani territory but a blueprint for how humanity can realign its values and economies to live in harmony with nature.

Ultimately, storytelling for activism serves as a conduit for creating a collective identity, one that is grounded in the stewardship of the Earth. The stories hold a mirror to society, showing how each of us is implicated in the environmental challenges we face and, more importantly, how each of us can be a part of the solution.

The stories from Nenquimo's and her fellow activists worldwide serve as an irreplaceable archive of human experience and environmental wisdom. These narratives do not just seek to change minds; they aim to transform hearts and compel us to recognize our shared humanity and our shared responsibility to protect the living planet that sustains us all.

As we consider the role of storytelling in fostering environmental consciousness and action, it is clear that the voices of those most affected by ecological degradation—the Indigenous populations, the local farmers, and the marginalized communities—must be amplified and placed at the center of the conversation. Their stories are not only accounts of survival against the odds; they are blueprints for a viable future, manifestos of hope, and calls to action that cannot be ignored if we are to forge a sustainable, equitable path forward for all Earth's inhabitants.

Chapter 21:
The Symphony of Sustainable Communities

As we draw our collective gaze toward the concept of sustainable communities, the analogy of a symphony is particularly fitting. Each community functions as a separate section of a grand orchestra, where diverse instruments play distinct roles, but when harmonized, they create a composition greater than the sum of its parts. Sustainable communities resonate with this orchestral harmony, achieving a balance between individual needs and the collective good, ensuring that the natural environment, economic vitality, and societal well-being are in concert. Creating these symphonies requires a profound understanding of how to harmonize interests to achieve a collective good where networks of interdependence support resilience in the face of challenges and envisioning real-world utopias becomes a shared vision rather than a solitary dream.

This chapter will resonate with the themes of interconnecting cultural, economic, and environmental aspects to orchestrate a future where sustainability isn't just an ideal but a tangible reality in which every community member plays a critical role in crafting continued success and harmony.

Harmonizing Interests for Collective Good

As we conduct the orchestra of sustainable communities, harmonizing interests for the collective good is the most intricate movement. The virtuoso performance we seek is one in which every individual plays their part yet blends seamlessly with the grander ensemble. The prime objective here is to align the diverse melodies of personal gain, communal prosperity, and environmental stewardship into a symphony that resonates across generations and borders.

The delicate task of harmonizing interests requires a profound understanding of how seemingly disparate goals can converge. It's pivotal to recognize sustainability isn't merely an environmental issue but a comprehensive framework that embraces economic viability, social equity, and cultural vitality. Economic growth, when orchestrated responsibly, can fuel the mechanisms for achieving both environmental and social objectives.

Communities are diverse, and so are their needs and aspirations. Harmonizing these interests involves cultivating a culture of dialogue and openness in which individuals and groups can voice their concerns and contribute to a collective vision. It's through this process that trust is built and a shared sense of purpose is established.

Furthermore, our conceptions of success and progress must be revisited. Traditionally, wealth accumulation has been a key indicator of prosperity. In the symphony of sustainable communities, however, success transcends material wealth to include sustainable living practices, protection of natural resources, and the health and happiness of all community members.

Environmental conservation efforts also play a crucial role in this endeavor. Individual and collective actions must prioritize the preservation of the planet as the cornerstone of all human activity. Sustainable resource management and safeguarding the Earth's bounty for both present and future populations becomes a shared interest.

In the realm of policymaking, harmonizing interests means implementing regulations that balance the need for economic development with environmental and social safeguarding. Policies need to be both incentive-based and regulatory—a blend that encourages innovation and holds parties accountable for their impact on the environment and society.

Businesses have a crucial part to play in this symphony. By adopting a triple bottom line approach, they can account for social and environmental performance alongside financial results. This approach encourages businesses to measure their success in terms of overall impact—a harmony among profits, people, and the planet.

Incentivizing sustainable practices is also instrumental. Rewards and recognition for businesses and individuals who demonstrate sustainable actions can drive wider adoption of such practices. This fosters a community environment where sustainability becomes the default tune to which all actions synchronize.

Education is the prelude to understanding and action. Educating community members about the interdependence of economic, social, and environmental systems enables them to understand the full consequences of their actions. This knowledge empowers people to make choices that align with the collective good.

Moreover, we need to embrace the diversity of ecological knowledge that exists within our global community. Indigenous and local knowledge systems have long understood the importance of living in harmony with nature and can guide modern practices.

Engagement and empowerment of all stakeholders is the key to harmonizing interests. When every voice is heard, when every person feels they have a stake in their community, sustainable practices become more than guidelines; they become the ethos of daily life.

This harmony is not static; it's a dynamic and ever-evolving process. As conditions change—whether through technological advancements, cultural shifts, or environmental pressures—our strategies and actions must adapt in concert.

Global collaboration amplifies the power of localized efforts. While sustainability starts at the community level, its effects ripple outward. Through partnerships and networks, best practices can be shared and scaled, echoing the adage that "we are stronger together."

Finally, perseverance and patience are vital. The path to harmonizing interests is not without discordance and setbacks. Yet, it's the commitment to staying the course, learning from missteps, and celebrating every victory that will see the symphony of sustainable communities flourish.

In conclusion, harmonizing interests for collective good is a multifaceted endeavor. It calls for a delicate balance of individual initiative and community cooperation, underpinned by a shared commitment to sustainable development. With concerted effort and an open heart, we can orchestrate a future that is truly sustainable for all.

Envisioning the Future: Utopias and Realities

In our quest for sustainability, we find ourselves at a crossroads between idealistic visions of utopia and the tangible realities we face. As we traverse the symphony of sustainable communities, we must pause to imagine what the future could hold. We must dream of worlds that not only survive but thrive—societies harmoniously in tune with the environment, where communities resonate with equity and well-being. The utopian visions of yesterday could be the blueprints for today's solutions, offering sketches of a world that align with our deepest ideals of harmony and balance.

At the heart of this vision lies the concept of eco-utopias, communities that have seamlessly integrated natural systems into their core functioning. Here, in these enclaves of theoretical perfection, humanity does not dominate nature but rather dances in step with it, creating a society where environmental stewardship is not merely an afterthought but the foundation upon which every decision is made.

The challenge, however, is to translate such aspirational visions into actionable reality. This requires a nuanced understanding that each community's path forward will be as unique as its cultural fabric, shaped by the diverse hands of its people and the particular rhythm of its local ecosystems. We must approach the future with plans that are not held captive by ideological rigidity but are adaptable and capable of evolving as they encounter the complexities of real-world applications.

Pragmatic utopianism suggests while we may not achieve an absolute ideal, we can strive for systems that are significantly better than those we currently have. This philosophy acknowledges the limitations of perfection but boldly insists upon relentless progress. It offers a future that, while perhaps not perfect, is infinitely better, more just, and more sustainable than the world we know today.

In envisioning such futures, technology stands out as a potent tool. Its judicious use can support sustainable energy, efficient urban design, and inclusive social structures, all aimed at reducing our ecological footprint. Sustainable technology, however, must be developed with a keen awareness of its cultural implications, ensuring it supports rather than disrupts the social and cultural tapestry it aims to preserve.

Art and culture, too, serve as gateways to imagining, and thus crafting, a more sustainable future. Through the power of storytelling, music, and visual arts, we can convey powerful messages and inspire communities to adopt more eco-friendly practices. The creative sectors act as the emotional compass of society, directing us toward a destination that aligns with our values and hopes.

Education remains a pivotal player in this orchestrated effort toward a sustainable utopia. It is through lifelong learning, rooted in diverse cultural contexts, that we can empower individuals and societies to participate in their local and global ecosystems with understanding and respect. The seeds of a sustainable future are planted in the classrooms and informal education forums where we nurture a new generation to think critically and act compassionately toward the environment.

Yet, as vital as these elements are to our envisioned future, we must remain cognizant of the economic structures that underpin our societies. A utopian future is not devoid of economic activity but is characterized by a green economy, one that is aware of the planet's finite resources and seeks to create prosperity without environmental compromise. Economic models must be recalibrated to account for the true cost of environmental degradation and to reward sustainable practices.

In this envisioned future, communities do not stand in isolation. Instead, they form a global mosaic, each contributing its unique cultural solutions to the collective challenges we face. This fabric of interconnected sustainability initiatives can generate a rich tapestry of knowledge and experience, a wellspring from which we all can draw. Here, the exchange of ideas and technology transcends borders, creating a solidarity that is both necessary and life-affirming in the face of global challenges.

It's also critical to confront the inevitable—the discrepancies between our dreams and reality. A utopian mindset could inadvertently set us on a path of disappointment if we hold unwaveringly to a static vision of perfection. The dynamic nature of societies, ecosystems, and the planet itself renders such an immutable destination unattainable. This does not mean, however, the pursuit is futile. Rather, it emphasizes the importance of resilience and adaptability as hallmarks of sustainable communities.

And what are the policies that will enable this envisioned future? We must design regulations and frameworks that are as adaptable and culturally sensitive as the communities they aim to serve. Multilateral cooperation is essential, with policy reflecting the shared responsibilities and varied capacities of nations and cultures worldwide. Policies must act as the guiding staves upon which the notes of our sustainable intentions can be played, resulting in an ensemble of solutions.

Moreover, as we envision the future, we must also critically assess the role each individual plays. Personal responsibility, married to collective action, sets the stage for deep transformations. Every choice, every action, echoes through the community—a reminder that sustainability is not only a societal goal but a personal journey. Cultivating a mindset of stewardship within every individual is as important as designing systemic changes.

Ultimately, sustainability is not a static end goal but a dynamic process, an ongoing practice of balance and re-evaluation. The utopias we envision serve as guideposts, ideals that propel us forward, but it is in the day-to-day realities, the incremental progress, we find true transformation. It is here, in the delicate balance between the dream of utopia and the pragmatism of reality, we craft a symphony of sustainable communities.

It's our collective imagination—a beautifully complex, ever-evolving vision of what can be—that continues to drive us toward a healthier planet and a more equitable society. These envisioned futures, a blend of utopian dreams and grounded realities, hold the promise of a world in which sustainable living is not merely a distant goal but an immediate, life-enhancing reality. They invite us not only to dream of a better tomorrow but to create it together, measure by measure, in the harmonious symphony of sustainable communities.

Chapter 22:
The Unfinished Quilt

As we reach the culmination of our journey, let us reflect on the metaphor of an unfinished quilt—a patchwork of diverse cultural fabrics, each piece essential, yet together still incomplete. It is a testament to the ongoing work necessary to achieve a sustainable and resilient world. Each chapter we've navigated has been akin to a stitch in this expansive tapestry, offering insights, theories, and practices crucial to understanding how we can advance sustainability in a world of rich cultural diversity.

Through the lens of the Anthropocene, we've come to understand the unprecedented impact humans have had on the planet. As each culture contributes its thread, it becomes increasingly clear that our shared future hinges on weaving these fragments into a coherent framework that honors our interconnectedness and interdependence.

Systems thinking has emerged as a vital needle in this process, punctuating the need to observe sustainability not merely as a linear progression but as a cyclical, inclusive loop. Drawing from chapters on education, technology, and policy, it is emphasized how intertwined knowledge, innovation, and governance are in cultivating an equitable future.

Our journey has also shined a light on the brilliance and wisdom of Indigenous communities, offering timeless insights that sync beautifully with modern sustainability initiatives. The deep-rooted understanding among these cultures signifies a harmonic engagement with our environment that we must continue to integrate into broader practices.

In moments of discord, it is the arts that often lead us back to harmony. Thus, weaving the rhythms of sustainability into the melodies of music and the strokes of a brush is not merely aesthetic; it's a poignant means of sparking dialogue and embodying change.

Green economics and the shift toward business models that embed cultural sustainability into their core represent another significant quilt patch. Recognizing prosperity and the health of our planet are intrinsically linked aligns economic practices more closely with cultural and ecological well-being.

Agriculture, as we have seen, serves as the root of resilience, with rural innovations demonstrating tradition and forward-thinking can coexist, offering a fertile ground for sustainable practice pebbles skipping across the water of modern challenges.

The fabric of social sustainability has revealed itself to be a rich, intertwining of social equity, health, and inclusion. These threads underscore true sustainability isn't only about the environment, but also involves fostering communities where every individual can thrive.

As we built the case for multilateral cooperation, the pattern for sustainability policy emerged—precision and artistry can yield policies sensitive to cultural nuances while still capably addressing global environmental challenges.

The stories of climate change and activism taught us the power of narrative and grassroots advocacy. These chapters served as reminders that every individual has a voice that can contribute to the chorus calling for environmental justice and action.

Technology and tradition, initially seeming at odds, were revealed as unlikely dance partners, each supporting the other in the quest for balance between progress and preservation. Tales of diverse cultural festivals then showed us how celebration and sustainability can joyously coalesce.

Culinary cultures, fashion, and fabrics each represent a realm in wich sustainability can be savored, worn, and embodied. These chapters fed our imagination with tastes and textures that serve as daily reminders of the impact of our choices.

In conclusion, we understand this quilt of ours is unfinished, not because our work is lacking, but because it is perpetually evolving. As with any worthwhile endeavor, the effort is never truly complete. With every new generation, unique patterns emerge, demanding integration into the quilt's greater design.

In closing, may we all embrace the unfinished state of our quilt, recognizing that it represents our shared commitment to a work in progress. As caretakers of this world our task is never done; perhaps that is as it should be. Instead, let us find inspiration in each new piece added and each repair made, and look forward to the endless possibilities as we continue to thread the needle toward a sustainable and resilient world.

Hold this metaphor close, and as you venture forward from the final page of this book, may your hands contribute to this tapestry with wisdom, care, and an unrelenting spirit of innovation that honors the past as it forges into the future.

Appendix:
Cultural Case Studies in Sustainability

I n this appendix, we explore a series of case studies that highlight the remarkable diversity of approaches to sustainability across cultures around the globe. From the ancient, terraced fields of the Philippines to the frozen tundras of the Arctic, local communities have crafted unique practices that have not only stood the test of time but now offer valuable insights into sustainable living for the rest of the world.

The case studies presented here testify to human ingenuity and the profound connection many cultures have maintained with their environment. They also shed light on how traditional knowledge systems and local solutions can contribute to global sustainability efforts and inspire actions that honor both cultural heritage and environmental stewardship.

Stories of Stewardship: Case Studies in Indigenous Sustainability

From the controlled burns and rotational cultivation practiced by Native American communities to the sustainable agroforestry methods of the Kayapó in the Brazilian Amazon, these case studies illustrate how Indigenous practices can inform and enrich contemporary sustainability efforts.

One significant example is that of the Waorani people in Ecuador, whose intimate knowledge of the Amazon rainforest has led to the protection of vast areas of biodiversity, vital not just for their survival but for the health of the planet. Likewise, the Ifugao Rice Terraces in the

Philippines demonstrate how ancient agricultural techniques can maintain soil integrity and water management while providing sustenance for entire communities.

Farther north, the Inuit hunting protocols present an excellent model of sustainable animal population management that ensures species thrive alongside human communities. Similarly, the Sámi reindeer herding traditions embody a delicate balance between human livelihoods and the preservation of ecological systems in the harsh conditions of the Arctic.

The Mbuti Pygmies' symbiotic relationship with the Ituri Forest in the Democratic Republic of the Congo and the Tuareg nomadic strategies in the Sahara Desert further reveal how survival and environmental management can be harmonious. Moreover, the ancient Qanat water systems in Iran exemplify ingenious engineering solutions designed to conserve and manage water in arid climates.

Not only are traditional agricultural practices—such as the multifaced terraced farming of the Hunza Valley in Pakistan and the agro-pastoral systems of the Brokpa in Bhutan—pertinent for local sustainability, they also offer frameworks that can be adapted to modern sustainable agriculture challenges. These stories, together with the methodologies of Indigenous Australians and Maori's guardianship approach to nature, expand our understanding of what constitutes sustainable practice and evoke a broader sense of ecological awareness.

The Native American Practices of Controlled Burns and Rotational Cultivation

As we delve deeper into the cross-cultural pursuit of sustainability, we encounter the wisdom inherent in the Native American land management practices. For millennia, these communities epitomized sustainability through practices like controlled burns and rotational cultivation, marrying an astute observation of ecological systems with practical stewardship.

Controlled burns—also known as cultural burns or fire stewardship—are an example of how Indigenous knowledge systems fostered biodiverse, resilient ecosystems. By intentionally setting low-intensity fires, Native Americans shaped their environments in a manner beneficial to both habitation and ecology. These burns cleared understory growth, returned nutrients to the soil, stimulated the germination of fire-adapted plant species, and managed pest populations. Furthermore, by reducing the fuel load, they dramatically decreased the risk and severity of potentially devastating wildfires.

In conjunction with fire, Indigenous cultures practiced rotational cultivation to prevent soil depletion and allow ecosystems to regenerate. This agroecological technique involved rotating crops and leaving fields fallow, thereby maintaining soil fertility and interrupting cycles of pests and diseases. It starkly contrasts the common monoculture practice that exhausts soil and leans heavily on chemical interventions.

These practices were not executed arbitrarily but were the result of careful observation and a deep understanding of local environmental cues. They fostered a dynamic equilibrium within ecosystems and underscored the principle that human activities should enhance, rather than diminish, biodiversity and ecological integrity. We find the Native American relationship with the land was characterized by reciprocity—a give-and-take—that is crucial to sustainability.

Today, revisiting these practices offers vital insights amidst climate crises and environmental degradation. Modern land management can benefit from incorporating these ancient yet scientifically sound practices to foster resilient landscapes capable of withstanding the storms of climate change. As we face these challenges, we must recognize the merit of learning from those who have long maintained the delicate balance of natural systems.

Embracing the Indigenous practice of controlled burning and rotational cultivation isn't merely an act of cultural preservation; it's an essential step toward sustainable land use that honors both history and science. It encourages us to reconsider our relationship with the environment, inspiring methods that support life in all its diversity. May we approach these ancient customs not only as echoes of the past but as living tenets of sustainability, evidence of a profound understanding of our interconnectedness with the Earth.

Kayapó Indigenous Agroforestry in the Brazilian Amazon

In the verdant expanses of the Brazilian Amazon, the Kayapó people have etched a living testament to the enduring wisdom of Indigenous stewardship. Their intricate agroforestry systems embody a profound understanding of ecological balance, offering insights into a symbiotic relationship between culture and land that is both ancient and indispensably modern.

Embarking on a journey through the Kayapó's agroforestry practices reveals a mosaic of cultivation that mirrors the biodiversity of the surrounding forest. This method of agroforestry—integrated by the Kayapó over generations—operates in harmony with the natural cycles of the forest, carving a middle path between conventional agriculture and the untamed wilderness. Unlike monocultures that deplete the soil and invite a cascade of ecological woes, Kayapó agroforestry nurtures a plethora of species, fostering resilience and sustaining a robust assortment of resources.

At its core, Kayapó agroforestry is a masterclass in agrobiodiversity. By alternating swaths of cultivated crops with natural forest, the ecosystem maintains its integrity, therefore supporting a vast array of plant and animal life. This careful management includes the cultivation of medicinal plants, fruits, and nuts, enriching the diet and health of the community while ensuring the land remains fertile and the forest canopy intact. Additionally, their practice of using "fire as a tool" to clear underbrush without harming mature trees has been preserved, showcasing a nuanced understanding of controlled ecological succession.

What stands as noteworthy in the tale of the Kayapó is not merely their conservation ethic but their proactive engagement with the land. It's a dynamic routine, a delicate dance of give-and-take with nature's rhythms, forging a living bond between the people and the Earth. Their agroforestry practices are a reflection of cultural beliefs that honor the interconnectedness of all life and underscore the inherent value of each being within the ecosystem.

For outsiders peering into the wisdom of the Kayapó, there lies potent potential for application in a world grappling with unsustainable agricultural practices. Emulation of these methods could pave the way for a revolution in how humanity sows and reaps the bounties of the Earth— a shift toward methods that heal rather than harm, replenish instead of ravage. The demonstration of these methods by the Kayapó people presents an invitation to the global community to reassess our relationship with the natural world.

The Kayapó have shown sustainability isn't a fad or a distant ideal. It's a way of life that has endured and thrived for centuries. With every plant they grow and every practice they observe, they issue a silent challenge to us: to learn, adapt, and adopt ways of living that promise a verdant future for generations to come.

Waorani People in Ecuador

The Waorani people of Ecuador, known for their deep connection to the rainforest and remarkable biocultural conservation strategies, offer us a profound lesson in the symbiosis of culture and environment. Living in the Amazon rainforest, a region with some of the highest biodiversity on the planet, the Waorani have cultivated a way of life that is inextricably linked to the forest they consider home. Their story is not just a chapter in the annals of Indigenous stewardship but is emblematic of the global struggle to maintain balance in the face of modern-day challenges.

Historically, the Waorani were known for their autonomy and warrior culture, which shielded them from colonial influences and the ensuing environmental degradation experienced by many other Indigenous groups. Their territory encompasses a mosaic of rainforests, rivers, and ancestral lands, which, to this day, they fiercely defend against external threats such as oil extraction and deforestation.

In terms of sustainability, the Waorani demonstrate a profound understanding of ecological processes, which they harness through practices such as traditional hunting, fishing, and small-scale agriculture geared toward subsistence rather than market-oriented production. These practices are informed by an intimate knowledge of their environment passed down through generations and embedded in a rich oral tradition that includes stories, songs, and rituals. Their livelihoods are, therefore, a conscious blend of conservation and utilization, minimizing their ecological footprint while sustaining their cultural heritage.

Recent conflicts have highlighted the Waorani's resilience and tenacity. In a groundbreaking legal case, they won a lawsuit against the Ecuadorian government, securing a legal precedent for Indigenous rights in the Amazon and halting oil drilling plans on their lands. This victory is a beacon of hope for Indigenous groups worldwide and underscores the importance of legal frameworks that recognize Indigenous territorial rights as a component of environmental stewardship and cultural preservation.

The challenges facing the Waorani are emblematic of the broader cultural and environmental threats that pervade Indigenous communities globally. However, their dedicated resistance and proactive strategies offer essential insights into how to harmonize traditional wisdom with contemporary conservation efforts, demonstrating sustainability is not just an environmental mandate but a cultural imperative as well.

By considering the Waorani as more than mere inhabitants of the rainforest but as its guardians, we are invited to rethink our own relationship with the natural world. Their legacy is a testament to the confluence where human culture and environmental custodianship meet. It reminds us every effort we make, whether local or global, is stitched into the broader tapestry of sustainability—a quilt that is ever-expanding, ever-evolving.

As policymakers, conservationists, and global citizens work to forge a path toward a more sustainable future, embracing the wisdom of the Waorani can lead to policies and practices that not only conserve the environment but enrich the cultural fabric of humanity. In this sense, the Waorani offer us a profound narrative: The most enduring solutions are those rooted in the timeless knowledge of the land and its people.

The Ifugao Rice Terraces: An Agricultural Marvel Rooted in Community and Sustainability

Reflect on the verdant stairways to the sky etched into the highlands of the Philippines—the Ifugao Rice Terraces. These terraces represent a supreme example of agriculture that transcends mere food production to embody the very essence of sustainability and community cohesion. For over two millennia, the Ifugao people have cultivated these slopes, ingeniously conforming to the rugged topography of the Cordillera mountain range. They have created not just a means of sustenance but a testament to human ingenuity and resilience.

It's in the wisdom of the Ifugao we find lessons so pertinent to our quest for sustainability. They've constructed a living system where water from the rainforests is channeled into an intricate irrigation network that nourishes each terrace, preserving both the natural watershed and their way of life. Their methods reflect a profound comprehension of the local ecology, integrating pest control and polyculture that minimize the need for artificial inputs. This close-knit bond between the community and their environment is a paragon for sustainable agriculture.

At the heart of the rice terrace's sustainability lies the "muyong," a privately-owned woodland that's managed with communal effort. The muyong system plays a critical role in preserving the terraces; it not only provides wood, but also safeguards against erosion, ensuring the soil remains fertile and retains water. The communal labor, or "bayanihan," demonstrates a cooperative spirit vital for the terraces' maintenance and a sense of shared responsibility among the individuals in the community.

Agriculturally, the Ifugao Rice Terraces embody a cycle of life that's as rhythmic as it is resilient. Seasonal patterns guide community activities, linking cultural practices to the rhythms of planting and harvest. This harmonious blend of culture and agriculture ensures traditions are passed on, keeping the community tightly knit together, even as external pressures mount.

Today, these terraces face challenges—from climate change to a labor shortage caused by youth migration. Yet, their existence continues to inspire. They remind us sustainable practices can create systems that thrive for generations, providing food, preserving culture, and fostering social solidarity.

In reflecting on the Ifugao Rice Terraces, we are impelled to consider the potency of cultural heritage in addressing contemporary sustainability challenges. They exemplify sustainability isn't a mere concept; it's a lived practice, dependent on a community's shared values and collective actions. It is an inspiring model for those searching for sustainable solutions grounded in the harmony between humankind and nature.

Indeed, the Ifugao offers a legacy on which we can build, showing us the future of sustainable agriculture may just lie in the treasured wisdom of the past. It's a reminder that sustainability doesn't have to be a grandiose abstraction—it's something tangible, achievable, and intrinsically linked to the communal spirit and respect for nature.

Inuit Hunting Protocols: Harmonizing Survival with Stewardship in the Arctic

In the heart of the Arctic, where the climate is as unforgiving as it is beautiful, the Inuit communities provide a profound example of sustainable hunting, deeply rooted in respect for nature and necessity. The glossy narrative that often accompanies conversations about sustainability can sometimes lead us to overlook the nuanced tapestry of traditional practices that have been fine-tuned over millennia to ensure survival in some of the harshest environments on Earth.

The Inuit way of life exemplifies a perfect harmony between the needs of the community and the rhythms of the natural world. Their hunting protocols, often misrepresented or misunderstood by distant onlookers, paint a vivid picture of judicious stewardship. The framework within which Inuit hunt is guided by principles of sustainability; these include not taking more than what is needed, respecting the animals and their habitats, and passing down knowledge through generations.

Sustainability is more than merely theoretical in the context of Inuit hunting. It's a lived, tangible practice that acknowledges the interdependence between humans and ecosystems. Inuit hunters employ methods that sustain animal populations and their environments, even improving habitat conditions for future generations. The methodical approach to hunting, known as "environmental stewardship," is one where hunters engage with their prey in a way that maintains the integrity of the ecosystem.

Central to Inuit hunting is the concept of "Avatimik Kamattiarniq," which can be translated as "environmental stewardship." It entails understanding the cumulative impacts of hunting and making informed decisions that promote the well-being of both the animal populations and the community. For example, the Inuit Qaujimajatuqangit, or IQ (traditional knowledge), is rich with guidelines such as the avoidance of wasteful practices, sharing of the catch within the community, and observing periods where hunting certain species is forbidden to allow for population recovery.

These practices underscore a fundamental belief in the reciprocity between humans and nature, an interconnectedness modern sustainability discourse sometimes neglects. In the face of climate change and industrial encroachment, these protocols are not just cultural relics but are vital contributions to global sustainability and biodiversity conservation efforts. Further, they offer valuable lessons in resource management that prioritize long-term ecosystem health over short-term gain.

It's important for policymakers and businesses to understand and respect these Indigenous systems and to recognize the depth of their environmental insight. By engaging with Inuit communities and incorporating their sustainable hunting protocols into broader conservation frameworks, we can foster a more inclusive approach to environmental stewardship.

The harmonization of survival with stewardship within Inuit cultures serves as a model for resilience and adaptability in the face of ecological adversity. Their protocols remind us that to achieve genuine sustainability, we must intertwine cultural knowledge with scientific understanding, crafting practices that nurture the planet even as they cater to human needs. As we move forward, embracing the wisdom of Inuit hunters becomes not just an act of respect for tradition but a crucial stride toward a sustainable future for all.

Sámi Reindeer Herding: A Deep-rooted Tradition of Coexistence and Adaptation in the Arctic

True sustainability hinges not only on modern technologies but equally on ancient customs that have stood the test of time. The Sámi people of the Arctic have demonstrated this poignantly through an intricate dance of adaptation and environmental stewardship that is epitomized in the traditional practice of reindeer herding. This practice, deeply engrained in Sámi culture, is a paradigm of coexistence with nature that offers valuable lessons for our global community as we grapple with the complexities of living sustainably in the twenty-first century.

Reindeer herding has been the cornerstone of Sámi livelihood for centuries. With an intuition honed by generations of symbiosis, the Sámi have developed a pastoral system that is remarkably attuned to the rhythms of the land and the lives of the reindeer. This system respects the migration patterns of the animals, following their natural movement across vast expanses of Arctic tundra and forested areas. It's a system that has fostered resilience both ecologically and culturally, enabling the reindeer and the Sámi to harness the sparse resources of the Arctic in a way that ensures their mutual survival.

However, what may seem like a simple tradition from the outside is, in actuality, a sophisticated model of sustainability. The reindeer provide the Sámi with food, materials for clothing and tools, and a vital connection to their heritage. In turn, the reindeer benefit from human protection and assistance in finding food during the harsh Arctic winters. This reciprocity is not static; it evolves with fluctuating climates and environmental conditions, demonstrating an extraordinary capacity for adaptation.

Current challenges—such as climate change, encroaching development, and competing land uses—put pressure on this delicate balance by altering the landscape and introducing new variables to an ancient equation. Yet, the Sámi demonstrate that cultural practices, when allowed to evolve organically, can be resilient enough to face modern threats. The community's relentless effort to adapt herding practices, involve youth, and engage in discourse with policymakers shows a proactive stance in preserving its culture and the sustainability it embodies.

It is critical for the global community to recognize the enduring value of such Indigenous traditions. By integrating lessons from the Sámi experience into broader sustainability discourses, we can better understand the nuances of ecological interdependence. The Sámi model teaches us sustainability is not merely about conservation but involves a profound connection to our environment, informed by the accrued wisdom and lived experiences of those who have thrived within it for millennia.

As we look to navigate the nexus of tradition and modernity, let the Sámi reindeer herding serve as an inspiration. Their story is a testament to the power of cultural resilience and the potential for human societies to live harmoniously with the Earth's ecosystems. It is clear that sustaining these practices is not merely a nod to our shared history; it is necessary for forging a sustainable future.

The Mbuti Pygmies of Ituri Forest in the Democratic Republic of the Congo

The Mbuti Pygmies of the Ituri Forest serve as a living testimony to a harmonious life within one of the world's most diverse ecosystems. Nestled in the heart of Africa, the Ituri Forest is a lush, biologically diverse haven and the homeland of the Mbuti people, who have forged a sustainable way of living intimately intertwined with their environment. Their practices offer us invaluable lessons on sustainability, showcasing how a community can thrive without compromising the health and vitality of its natural surroundings.

Tucked away from the rapid urbanization and technological advancements of modern society, the Mbuti have managed to preserve a lifestyle many would regard as a blueprint for ecological conservation and cultural endurance. This community prioritizes the forest not just as a resource for survival but as a sacred entity that commands respect and careful stewardship. This ethos is imbued in all aspects of their culture, from hunting to habitat management.

The Mbuti operate within a framework of traditional ecological knowledge, which is passed down through generations. Hunting and gathering practices are carried out with precision and care, ensuring wildlife populations remain stable and the forest can continue to flourish. The notion of taking only what is needed—a principle that contrasts starkly with the excesses of consumption-driven cultures—is deeply ingrained in their way of life. In the Ituri Forest, the Mbuti Pygmies demonstrate a profound understanding that human well-being is directly linked to the health of the ecosystem.

What's more, Mbuti social structures and norms enforce a symbiotic relationship with nature. Decisions are often made collectively, fostering a sense of social cohesion and shared responsibility for the environment on which they depend. This level of community engagement with natural resource management exemplifies systems thinking—a vital component for sustainable communities globally.

The challenges the Mbuti face, including deforestation, mining, and civil unrest within the Democratic Republic of the Congo, threaten their way of life and the biological integrity of the Ituri Forest. Yet, their resilience and the adaptive nature of their culture are litmus tests for the possibilities of living sustainably within a rapidly changing world. As policymakers and global actors search for pathways to a sustainable future, they can draw inspiration from the Mbuti's resilient model of coexistence with the natural world.

In conclusion, the Mbuti Pygmies of Ituri Forest stand as more than just a case study in sustainability; they are a beacon of hope and a reminder that living sustainably is not a newfangled concept but one deeply rooted in Indigenous wisdom. Embracing their values and practices could help steer our global community toward a future that honors the interconnectedness of life and the necessity of preserving the delicate balance between humans and nature.

Tuareg people of the Sahara Desert

In the heart of the Sahara, the world's largest hot desert, reside the Tuareg people—an Indigenous nomadic community whose remarkable way of life has been intricately woven into the very fabric of the arid landscape they inhabit. The Tuareg are known as "the blue people" for the indigo dye of their garments, which, like their culture, is resilient in the face of the relentless sun and wind-swept sands. Their story offers a profound lesson in sustainability, resilience, and the art of thriving under seemingly inhospitable conditions.

For centuries, the Tuareg have navigated a delicate balance with their environment, embodying sustainability long before the term entered the lexicon of global discourse. Their nomadic lifestyle, guided by the seasons and the search for water and pasture, is a testament to a sophisticated understanding of their ecosystem. This deep-rooted knowledge is often disregarded in the modern dialogue on climate adaptation, yet it holds the keys to sustainable living in extreme climates.

The social fabric of the Tuareg is tightly knit. Their communities are bound by a code of honor, ancestry, and a governance system that eschews centralization in favor of clan-based decision-making. This structure underscores the importance of social equity and the collective stewardship of resources, an approach modern societies could learn from when confronting environmental challenges and pursuing sustainable futures.

In a landscape where resources are scarce, the Tuareg have mastered the art of using what is available sustainably. Their traditional use of local materials to construct homes and tools demonstrates a circular economy that minimizes waste and maximizes utility. Similarly, their herding and agricultural practices are acutely adapted to the ebb and flow of the Sahara's harsh climate to ensure food security and the balance of the region's biodiversity.

Their knowledge of the Indigenous flora and fauna of the Sahara is not merely a matter of survival but represents a harmonious interplay with the natural world—characteristics imperative for global sustainability efforts. With their ability to traverse vast, unforgiving terrains, the Tuareg have cultivated a resilience that enables them not just to survive, but also to maintain their cultural integrity while the climate around them shifts.

However, as is the case with numerous Indigenous communities worldwide, the Tuareg face mounting challenges. Climate change, political instability, and external pressures on the environment pose a significant threat to their way of life. It is crucial for policymakers and sustainability advocates to recognize the value of Indigenous knowledge systems and ensure the Tuareg have a voice in the dialogue about the future of our planet.

By understanding and respecting the symbiotic relationship the Tuareg people have with the Sahara Desert, there is an opportunity to learn from their deep connection to the Earth. In an era marked by disconnection and environmental disregard, their life philosophy and practices can inspire and inform broader strategies for living sustainably within one's means and respecting the non-negotiable limits set by nature.

The Tuareg people, with their centuries-old wisdom, offer a blueprint for resilience and adaptation that transcends the harshness of the Sahara. They teach us sustainability is not simply a technical challenge to be engineered but a deep cultural practice rooted in respect for the Earth and an unwavering solidarity with all its inhabitants. Theirs is a narrative that deserves amplification, a cultural thread in the broader tapestry of sustainability that we all must weave together.

The Qanat System of Iran: Ancient Aqueducts for Sustainable Water Management

In the arid expanse of the Iranian Plateau, where the unforgiving sun reignites an ancient thirst within the sands, lies a marvel of engineering so profound in its simplicity and effectiveness that it has withstood the test of millennia. The qanat system, a network of gently sloping underground channels, has been an enduring testimony to human ingenuity in the face of environmental adversity. This intricate grid of waterways transcends mere irrigation techniques; it embodies a sustainable philosophy, merging ecologically sound practices with equitable distribution. The Qanat of Iran isn't just about water; it's a living narrative of culture, community, and wisdom buried in age-old traditions.

The qanat, also known as kariz, is designed to tap into the mountain aquifers at the head of a valley, gently guiding the water through underground tunnels to the farmlands and settlements in the dry plains. Using gravity and skilled craftsmanship, the ancient Persians harnessed a stable water flow despite the harsh surface environments. By avoiding evaporation losses typical in open canals, the qanat system is incredibly efficient; each precious drop of water, the lifeblood for settlements in the otherwise inhospitable terrain, is conserved.

At its core, the qanat is not merely a triumph over aridity; it is a testament to the collective commitment and the deep understanding that sustainability is a community endeavor rather than an individual conquest. The qanat brings life, but it also requires life through a continual investment of care and labor. Its upkeep and the equitable management of water define an entire social structure, wherein water masters called "mirab" orchestrate the distribution of water, ensuring the balance between human needs and natural limits.

The lessons from Iran's ancient aqueducts resonate with profound pertinence today. These subterranean arteries of life empower us to think about our contemporary challenges through the lens of sustainability. They beckon us to explore how we might address our water crises globally, not through further straining our overextended rivers and aquifers but by embracing a measured, communal strategy rooted in conservation and balance. Entrepreneurs, policymakers, and communities alike can draw from this well of wisdom, recognizing true sustainability is a fusion of cultural appreciation, environmental stewardship, and social justice.

In an era where resilience and adaptability are keystones for survival, the qanat system stands as an inspiring reflection of what can be achieved. It compels us to ponder upon our interdependence with nature and with each other, urging a recalibration of our relationship with the Earth's resources. The Iranian qanats are not just relics; they are beacons, illuminating a path forward for a world yearning for sustainable solutions.

The Pamiri Houses and Traditional Agriculture of the Gorno-Badakhshan Region, Tajikistan

In the high-altitude embrace of the Pamir Mountains, the Indigenous people of the Gorno-Badakhshan Autonomous Region in Tajikistan have curated a lifestyle profoundly attuned to the pillars of sustainability—traditional Pamiri houses and agriculture are testaments to this. These homes, with their ingenious architectural elements, are not only physical structures but embodiments of cultural resilience and environmental wisdom.

Pamiri houses, often called "Chid," are built using local materials such as stone, wood, and mud, merging seamlessly with the rugged landscape. Their design embodies sustainability: the five-pillared central room represents the five elements of Ismaili belief—wood, fire, earth, metal, and water—and reflects the symbiosis of spiritual and material lives. These dwellings are designed to withstand the harsh climate, with thick walls providing insulation and large windows capturing the warmth of the sun, demonstrating an intuitive understanding of passive solar heating long before it became a tenet of sustainable architecture.

Complementing the ingenuity of Pamiri architecture, traditional agriculture in Gorno-Badakhshan has flourished through a remarkable adaptation to mountainous terrain and scarce arable land. Terraced farming contours along the slopes, maximizes agricultural space, prevents soil erosion, optimizes water use, and enables micro-climate cultivation. The intercropping of plants such as barley, wheat, and various fruits creates a biodiverse palette resilient to pests and diseases while sustaining soil fertility.

Additionally, these farming practices embody the community's respect for the land. Crop rotation and fallowing of fields are not new trends but ancient wisdom that has maintained soil health for generations. Water, a scarce resource, is managed through intricate irrigation systems known as "aarband," which distribute meltwater from glaciers and springs equitably among community members. This system reflects the community's deep understanding of the hydrological cycle and the necessity of shared resource management.

The cultural landscape of Gorno-Badakhshan, with its traditional Pamiri houses and agricultural practices, reveals a harmonious interplay between human and natural systems. As the planet confronts the pressures of climate change and resource depletion, these enduring practices offer valuable insights. They exemplify how cultural traditions, when underpinned by respect for the natural world, can lead to sustainable living solutions that are both innovative and deeply rooted in place.

For readers seeking to weave sustainability into the fabric of their own lives and communities, the lessons from Gorno-Badakhshan's culture provide a compelling narrative. They encourage a reevaluation of how we construct our homes, manage our resources, and cultivate our food, urging us to take inspiration from those who have thrived in harmony with Earth's rhythms for centuries.

The Hunza Valley's Agroforestry and Terraced Farming, Pakistan

Embedded within the rugged landscape of northern Pakistan lies the idyllic Hunza Valley, a testament to the enduring relationship between culture and sustainable agriculture. Here, communities have long practiced terraced farming, an intricate choreography between human toil and nature's bounty that balances on the precarious slopes of the Karakoram Range. This mosaic of terraces serves as a canvas where the art of agroforestry takes on a vivid form, blossoming against the backdrop of mountain peaks.

Straddling the banks of the Hunza River, the terraced gardens are marvels of engineering crafted over generations. They are a testament to an ancestral wisdom that intricately links the land's productivity with its preservation. It's in this confluence where one finds the essence of sustainability: a balance between yielding harvests and maintaining the land's vitality.

The terraced farming system in Hunza ingeniously conserves soil, decreasing erosion by rain and the flowing river. Each level step of the terraces traps water, making irrigation efficient and minimizing waste. Furthermore, the agroforestry practices here blend cultivation with the maintenance of trees that provide fruits, timber, and fodder while also stabilizing the terraces and contributing to soil fertility through leaf litter.

What truly resonates is how the Hunza model of agriculture encapsulates the symbiosis between people and their environment. These practices reflect a profound knowledge of local ecology and an understanding that each species planted must serve multiple purposes. They must not only thrive in the harsh mountainous climate, but also support the needs of the community for food, fuel, and fodder, while safeguarding the environment.

The wisdom of the Hunza is not merely to be revered but to be understood and integrated. Their approach offers valuable lessons in sustainable resource management, demonstrating agricultural practices can be tailored to the unique features of the landscape to create a resilient and self-supporting system. Such insights hold immense potential for informing sustainable rural development in similar environments worldwide.

Acknowledging the practices of the Hunza Valley is not a romanticization of tradition but an acknowledgment of their relevance in contemporary sustainability discourse. It embodies the unequivocal notion that there are ancient threads in the fabric of modern sustainability that are robust, adaptable, and teeming with potential.

The resilient systems of agroforestry and terraced farming in Hunza offer not just a lifeline to the communities that nurture them, but also shape a discourse of sustainable agriculture that resonates globally. In a world grappling with the adverse effects of climate change and food insecurity, the practices of Hunza stand as a beacon, illuminating a path that is both ancient and urgently relevant for the future of our collective well-being on this planet.

Brokpa People of Bhutan: Ancestral Agro-Pastoralism in Harmony with Nature

As we voyage through the intricate weave of cultural threads contributing to the tapestry of global sustainability, one cannot help but stand in admiration of the Brokpa tribe of Bhutan. Tucked within the eastern Himalayas, this group of roughly 5,000 individuals is a beacon of agro-pastoral practices that have stood the test of time, persisting harmoniously with the natural world. The Brokpa have diligently maintained a resilient lifestyle that is as much an embodiment of their cultural identity as it is a testament to their sustainable agriculture and pastoralism.

Embedded within their culture is a profound reverence for their environment. Agro-pastoralism among the Brokpa involves an intricate balance of crop cultivation and animal husbandry, crafted to suit the high-altitude terrains they call home. They grow hardy crops such as barley, maize, and potatoes, and they rear livestock, including yaks and sheep, which are not just sources of food, but also central to their economic livelihood and cultural ceremonies.

Remarkably, the Brokpa people's sustainability goes beyond mere subsistence and enters the realm of spiritual connection with their surroundings. The yak, indispensable to their lifestyle, provides dairy products, meat, and wool for clothing—whilst also serving as a symbol of prosperity and well-being. Their sustainable agricultural practices are mindfully designed to rotate crops and rest fields, ensuring soil fertility is preserved. Even in challenging high-altitude conditions, these practices ensure they live within the carrying capacity of their environment.

Intimately entwined with their surroundings, the Brokpa have mastered the art of listening to the land, responding to the subtle cues of nature, and adapting their methods accordingly. This adaptive capacity is a cornerstone of resilience within their community, enabling them to flourish despite the uncertainties brought by a changing climate. The community's collective wisdom in predicting weather patterns and their management of natural resources has allowed them to sustain not just their way of life but the biodiversity that thrives alongside them.

As the world grapples with finding sustainable solutions, the Brokpa community's practices offer profound insights into the power of cultural traditions in environmental stewardship. They are living proof that Indigenous knowledge systems, when respected and preserved, can guide societies toward a harmonious coexistence with nature.

In essence, the Brokpa people demonstrate sustainability is not just about innovation; it is also about the preservation and continuation of ancestral knowledge that has been refined over centuries. Their legacy is a clarion call for the global community to witness and learn from, to recognize there lies immense value in the diverse ways humanity understands and interacts with our Earth.

The Brokpa's way of life serves as a powerful model for embedding sustainability into the cultural fabric of society—illustrating that when we nurture our ties to the land and to each other, a more sustainable and resilient world is not just possible, but already in existence within the heart of these communities.

Glossary of Terms in *Our Changing World*

Throughout our conversation on *Our Changing World* and the multifaceted dimensions of sustainability, we've encountered a lexicon that is as varied as it is vital – a language shaping our understanding of not just where we stand, but also where we must go. This glossary serves as a compass to navigate the terrain of sustainability terms that appear throughout the book, ensuring clarity and fostering a deeper appreciation for the urgency and beauty in this journey toward a sustainable future.

Acculturation
Acculturation is a complex and multifaceted process where individuals or groups from one culture engage in continuous and direct interaction with another culture. This process leads to the adoption of new values, behaviors, and practices from the encountered culture.

Adaptation
The process through which individuals, communities, and ecosystems adjust to changes in their environment to mitigate harms or exploit beneficial opportunities. It is a cornerstone for resilience in the face of climate change, adjusting to its inevitable impacts.

Albedo
A measure of how much light that hits a surface is reflected without being absorbed. Surfaces with high albedo, such as ice and snow, can reflect more sunlight and affect climate by keeping temperatures cooler, whereas surfaces with low albedo, like forests or oceans, absorb more sunlight and can contribute to warming.

Animism
A belief system that posits that all objects, places, and creatures possess a distinct spiritual essence. In sustainability, this perspective can influence environmental policies and conservation efforts, emphasizing the intrinsic value of nature and the need to respect and preserve the spiritual integrity of the natural world.

Anthropocene
A proposed epoch marking the significant global impact of human activity on the Earth's geology and ecosystems, including climate change and biodiversity loss

Agroforestry
The practice of integrating trees and shrubs into crop and livestock farming systems to increase biodiversity, improve soil health, and enhance ecosystem services

Biodiversity
The variety within and among all species of plants, animals, and micro-organisms and the ecosystems of which they are part. This includes diversity within species, between species, and of ecosystems, forming the web of life of which humans are an integral part and upon which they fully depend.

Carbon Footprint
A measure of the total amount of greenhouse gases produced to directly and indirectly support human activities, typically expressed in equivalent tons of carbon dioxide (CO_2). Reducing our carbon footprint is key to combating climate change.

Circular Economy
An economic system aimed at minimizing waste and making the most of resources. This regenerative system aims to close the gap between production and natural ecosystem cycles—a stark contrast to the

traditional linear economy, which has a 'take, make, dispose' model of production.

Climate Change
A long-term change in the average weather patterns that have come to define Earth's local, regional, and global climates. These changes have a broad range of observed effects that are synonymous with the term global warming.

Conservation
The protection, preservation, management, or restoration of natural environments and the ecological communities that inhabit them. Conservation is a means to ensure that nature will be around for future generations to enjoy and also recognizes the integral role nature plays in providing ecosystem services.

Corporate Social Responsibility (CSR)
A framework for businesses to voluntarily integrate social and environmental considerations into their operations and interactions with stakeholders. It's a commitment to manage the economic, social, and environmental impacts of a company's operations responsibly and in line with public expectations.

Cultural Capital
The collection of symbolic elements such as skills, tastes, posture, clothing, mannerisms, material belongings, and credentials that one acquires through being part of a particular social class.

Cultural Homogenization
The process by which local cultures are assimilated and eroded by dominant cultures, often as a result of globalization. This can lead to a loss of cultural diversity and the unique knowledge and practices that contribute to sustainability and resilience in various environmental contexts.

Cultural Sensitivity
Awareness and respect for cultural differences and the willingness to understand, communicate with, and effectively interact with people across cultures.

Cultural Sustainability
Maintaining and evolving cultural beliefs, practices, and heritage as part of a community's overall sustainability goals, with respect for diversity and tradition.

Cultural Tipping Points
Moments when a cultural norm or practice reaches a threshold and spreads rapidly, which can have significant implications for social sustainability.

Eco-Friendly Apparel
Clothing made from organic or recycled materials, using processes that minimize the environmental footprint during production, distribution, and disposal.

Eco-efficiency
Eco-efficiency is achieved by delivering competitively priced goods and services that satisfy human needs and bring quality of life, while progressively reducing ecological impacts and resource intensity throughout the life-cycle, to a level at least in line with the Earth's estimated carrying capacity.

Ecological Footprint
A measure of how much area of biologically productive land and water an individual, population, or activity requires to produce all the resources it consumes and to absorb the waste it generates.

Ecosystem Services
The many and varied benefits to humans that are provided by the natural environment and from healthy ecosystems. These include, but are not

limited to, provisioning, regulating, cultural, and supporting services that directly or indirectly benefit human well-being.

Environmental, Social, and Governance (ESG)
Environmental, Social, and Governance (ESG) criteria are a set of standards for a company's operations that investors use to screen potential investments. Environmental criteria consider how a company safeguards the environment; social criteria examine how it manages relationships with employees, suppliers, customers, and communities; governance deals with a company's leadership, executive pay, audits, internal controls, and shareholder rights. ESG examines a company's non-financial materiality that may influence investor engagement.

Environmental Stewardship
The responsible management and care of the environment and natural resources with an emphasis on preserving and enhancing biodiversity and ecological integrity.

Epistemological Diversity
Recognition of the existence of multiple ways of knowing and understanding the world, which can be influenced by culture, language, and personal experience.

Ethical Consumerism
The practice of purchasing products and services that are produced ethically, considering the labor conditions, environmental impact, and animal welfare. Ethical consumerism encourages sustainable production practices and corporate social responsibility, influencing market trends toward more sustainable options.

Feedback Loop
A system where the output of a process is used as an input, leading to further output that may amplify (positive feedback) or diminish (negative feedback) the process.

Geopolitical Tensions
Political tensions influenced by geographic factors, often related to resource conflicts or environmental impacts.

Green Economics
An economic framework that takes into account ecological and social costs, promotes sustainability, and values the well-being of both the environment and society.

Grassroots Movements
Local, community-driven movements that grow to influence larger populations and policies, often associated with sustainability.

Holistic Thinking
An approach that considers the interconnectedness and complexity of systems, crucial for addressing sustainability challenges.

Indigenous Knowledge
Local knowledge unique to a culture, often contrasting with global knowledge systems and important for sustainability.

Indigenous Land Rights
The recognition of Indigenous peoples' rights to their traditional lands and resources, essential for cultural preservation and sustainable practices.

Interconnectedness
The recognition of the dependence of all life forms and ecosystems on each other, leading to an understanding that actions taken in one area can have global implications.

Intergenerational Equity
The concept of fairness or justice in relationships between the present and future generations, particularly in terms of resource allocation.

Isolationism
A policy or doctrine of isolating one's country from the affairs of other nations by declining to enter into alliances, foreign economic commitments, international agreements, etc., seeking to devote the entire efforts of one's country to its own advancement and remain at peace by avoiding foreign entanglements and responsibilities. In sustainability, isolationism can hinder global collaborative efforts needed to address worldwide environmental and social challenges.

Moral Imperatives
Principles or ethical considerations that compel individuals or societies to act in accordance with what is considered right and just.

Multicultural Curriculum
Educational syllabi that incorporate diverse cultural perspectives and content, thereby promoting inclusivity and understanding among students of different backgrounds.

Regenerative Design
Regenerative Design is a process-oriented approach to design. The goal is to develop systems that are capable of regenerating or restoring their own sources of energy and materials, thus creating sustainable patterns of consumption and production.

Renewable Energy
Energy from sources that are not depleted when used, such as wind or solar power, which are essential for sustainable development.

Resilience
The capacity of a system, community, or society potentially exposed to hazards to adapt, by resisting or changing in order to reach and maintain an acceptable level of functioning and structure. This is determined by the degree to which the social system is capable of organizing itself to increase its capacity for learning from past disasters for better future protection and to improve risk reduction measures.

Social Cohesion
The willingness of members of a society to cooperate with each other in order to survive and prosper, which can be essential for sustainable development.

Social Resilience
The ability of a community to withstand external shocks and stresses as a result of social capital and community resources.

Social Equity
The fair and just treatment of all individuals within society, ensuring equal access to opportunities and resources, and the protection from discrimination.

Societal Equilibrium
A state of balance in a society or ecosystem, where the social or natural structure is maintained over time, often through sustainable practices.

Socioeconomic Resilience
The ability of a social and economic system to recover from shocks and stresses, such as economic crises or natural disasters.

Sustainability
The ability to meet the needs of the present without compromising the ability of future generations to meet their own needs. Sustainability is often broken into three pillars: environmental, economic, and social, also known informally as planet, profit, and people.

Sustainable Agriculture
Farming that meets the needs of the present without compromising the ability of future generations to meet their own needs, typically involving environmentally friendly practices.

Sustainable Development
Development that meets the needs of the present without compromising the ability of future generations to meet their own needs, encompassing a balance between environmental, economic, and social goals.

Systems Thinking
A holistic approach to analysis that focuses on the way a system's parts interrelate and how systems work over time within the context of larger systems.

Tipping Point
A critical threshold at which a small change or influence can lead to a significant and often irreversible effect on a system. In sustainability, tipping points are crucial in the context of climate change and biodiversity, where they represent points beyond which systems may not recover, leading to drastic changes in the environment.

Traditional Ecological Knowledge (TEK)
A cumulative body of knowledge, practice, and belief, evolving by adaptive processes and handed down through generations by cultural transmission, about the relationship of living beings (including humans) with one another and with their environment.

Tragedy of the Commons
A situation in a shared-resource system where individual users acting independently according to their own self-interest behave contrary to the common good of all users by depleting or spoiling that resource through their collective action.

Zero-Carbon
Referring to an operation or activity that releases no carbon dioxide into the atmosphere. This ambitious goal can be approached by reducing emissions and implementing carbon offset schemes that compensate for any emissions that are produced. Zero-carbon is a guiding star in the journey toward climate neutrality.

References

Acabado, S. (2009). A Bayesian approach to Dating agricultural terraces: A case from the Philippines. Antiquity, 83(321), 801-814.

Acabado, S. (2019). The Ifugao Agricultural Landscapes: Complementary Systems and the Intensification Debate. Journal of Southeast Asian Studies, 50(3), 423-443.

Acosta, A., Calle, A., & Suárez, D. C. (2020). Redefining socio-ecological systems: The case of the Indigenous territoriality in the Amazon basin. Environment, Development and Sustainability, 22(5), 4521-4544.

Adger, W. N. (2006). Vulnerability. Global Environmental Change, 16(3), 268-281.

Adger, W. N., Barnett, J., Brown, K., Marshall, N., & O'Brien, K. (2013). Cultural dimensions of climate change impacts and adaptation. Nature Climate Change, 3(2), 112-117.

Agyeman, J., & Carmin, J. (2011). Environmental Inequity in Urban America and Community-Based Environmental Organizations. Environmental Science & Policy, 9(5), 564-574.

Agyeman, J., Bullard, R. D., & Evans, B. (Eds.). (2003). Just sustainabilities: Development in an unequal world. MIT Press.

Alcorn, J. B., & Toledo, V. M. (1998). Resilient resource management in Mexico's forest ecosystems: The contribution of property rights. In F. Berkes & C. Folke (Eds.), Linking Social and Ecological Systems: Management Practices and Social Mechanisms for Building Resilience (pp. 216-249). Cambridge University Press.

Altieri, M. A., & Nicholls, C. I. (2012). Agroecology scaling up for food sovereignty and resiliency. In E. Lichtfouse (Ed.), Sustainable Agriculture Reviews (Vol. 11, pp. 1-29). Springer, Dordrecht. https://doi.org/10.1007/978-94-007-5449-2_1

Altieri, M. A., Funes-Monzote, F. R., & Petersen, P. (2015). Agroecologically efficient agricultural systems for smallholder farmers: Contributions to food sovereignty. Agronomy for Sustainable Development, 35(1), 1-13.

Anderson, A. (2015). Engaging with indigenous knowledge systems in sustainable natural resources management. Progress in Physical Geography: Earth and Environment, 39(5), 572-590.

Andersson, P., & Bateman, T. (2017). Reimagining sustainability in precarious times. Education + Training, 59(2), 184-198.

ARC. (2020). Faith Commitments for a Living Planet. Alliance of Religions and Conservation.

Bailey, R. C., Head, G., Jenike, M., Owen, B., Rechtman, R., & Zechenter, E. (1999). Hunting and gathering in tropical rain forest: Is it possible? American Anthropologist, 101(1), 82-93.

Barthel, S., Folke, C., & Colding, J. (2010). Social-Ecological Memory in Urban Gardens—Retaining the Capacity for Management of Ecosystem Services. Global Environmental Change, 20(2), 255-265.

Beaumont, P. (1989). Qanat Systems in Iran: An Historical and Architectural Study. Bulletin of the School of Oriental and African Studies, 52(3), 550-561.

Benessia, A., Funtowicz, S., Bradshaw, G., Ferri, F., Ráez-Luna, E., & Medina, C. P. (2012). Hybridizing sustainability: Towards a new praxis for the present human predicament. Sustainability Science, 7(S1), 75-89.

Berkes, F. (2009). Indigenous ways of knowing and the study of environmental change. Journal of the Royal Society of New Zealand, 39(4), 151-156.

Berkes, F. (2017). Sacred Ecology. Routledge.

Berkes, F. (2018). Sacred Ecology. Routledge.

Berkes, F., & Folke, C. (1998) Linking social and ecological systems for resilience and sustainability. Linking social and ecological systems: management practices and social mechanisms for building resilience. Cambridge University Press, Cambridge, UK.

Blench, R. (2001). 'You can't go home again': Pastoralism in the new millennium. Overseas Development Institute.

Bodansky, D. (2010). The Art and Craft of International Environmental Law. Harvard University Press.

Bogoslowski, T., Szymkowiak, M., & Lipowski, M. (2019). Dietary habits and body condition in Nenets reindeer herders on the Russian tundra. Arctic Anthropology, 56(1), 1-17.

Bourdieu, P. (1986). The forms of capital. In J. G. Richardson (Ed.), Handbook of Theory and Research for the Sociology of Education (pp. 241-258). Greenwood.

Boykoff, M. T. (2011). Who Speaks for the Climate? Making Sense of Media Reporting on Climate Change. Cambridge University Press.

Brosius, J. P. (1997). Endangered forest, endangered people: Environmentalist representations of indigenous knowledge. Human Ecology, 25(1), 47-69.

Brundtland, G. H., Khalid, M., Agnelli, S., Al-Athel, S., Chidzero, B., Fadika, L., ... & Singh, M. (1987). Our common future ('brundtland report').

Bulkeley, H., & Newell, P. (2010). Governing climate change. Routledge.

Caporaso, J. A., & Keeler, J. T. S. (1995). The European Union and regional integration theory. In C. Rhodes & S. Mazey (Eds.), The state of the European Union, 3, 29-62.

Capra, F. (1996). The Web of Life: A New Scientific Understanding of Living Systems. Anchor Books.

Crate, S. A. (2011). Climate Change and Ice Festivals: Cultural Sustainability in the Andes. Ethnology, 50(2), 161–178.

Curtis, D. J. (2009). Creating inspiration: The role of the arts in creating empathy for ecological restoration. Ecological Management & Restoration, 10(3), 174-184.

Daly, H. E., & Farley, J. (2011). Ecological Economics, Second Edition: Principles and Applications. Island Press.

Denton, F. (2002). Climate change vulnerability, impacts, and adaptation: Why does gender matter? Gender & Development, 10(2), 10-20.

Diaz, R. (2020). Technical training for sustainable development: Evolving competencies in a changing world. International Journal of Technical Education and Training, 2(1), 1-13.

Dove, M. R., & Carpenter, C. (2008). Environmental Anthropology: A Historical Reader. Blackwell Publishing.

Doyle, C. (2015). Indigenous Peoples' Rights to Lands, Territories, and Resources. International Work Group for Indigenous Affairs.

Durie, M. (1998). Whaiora: Māori Health Development. Oxford University Press.

EcoIslam. (2020). Eco-lifestyle: Footprint Before Handprint. British Muslim Environmental Network.

Edenhofer, O., Pichs-Madruga, R., Sokona, Y., Farahani, E., Kadner, S., Seyboth, K., ... & Minx, J. C. (Eds.). (2019). Climate Change 2014: Mitigation of Climate Change. Cambridge University Press.

EPA. (2021). Environmental Justice. Environmental Protection Agency. Retrieved from https://www.epa.gov/environmentaljustice

Fletcher, K. (2014). Sustainable Fashion and Textiles: Design Journeys. Routledge.

Fletcher, K., & Grose, L. (2012). Fashion & Sustainability: Design for Change. Laurence King Publishing.

Folke, C., Carpenter, S., Walker, B., Scheffer, M., Elmqvist, T., Gunderson, L., & Holling, C. S. (2004). Regime shifts, resilience, and biodiversity in ecosystem management. Annual Review of Ecology, Evolution, and Systematics, 35, 557-581.

Gardner, G. T. (2003). Invoking the Spirit: Religion and Spirituality in the Quest for a Sustainable World. Worldwatch Institute.

Garnett, S. T., Burgess, N. D., Fa, J. E., FernÃ¡ndez-Llamazares, MolnÃ¡r, Z., Robinson, C. J., ... & Collier, N. F. (2018). A spatial overview of the global importance of Indigenous lands for conservation. Nature Sustainability, 1(7), 369-374.

Gautam, R. (2013). The road to longevity: Cultural and biological perspectives. In L. J. K. G. Thorpe (Ed.), Culture, Biology, and Anthropological Demography (pp. 74-92). Cambridge University Press.

Giannachi, G., & Stewart, N. (2005). Performing Nature: Explorations in Ecology and the Arts. Peter Lang.

Gibson, C., Carr, C., & Warren, A. (2019). Making music, making waste: Waste and resource recovery in creative economy sectors. Geoforum, 102, 142-151.

Goldstein, B. E., Wessells, A. T., Lejano, R., & Butler, W. (2020). Narrative strategies for the artful management of environmental challenges. Environmental Management, 66(1), 1-13.

Green Pilgrimage Network. (2021). Green Pilgrimage Handbook. Alliance of Religions and Conservation.

Hale, T. (2016). "All hands on deck": The Paris Agreement and nonstate climate action. Global Environmental Politics, 16(3), 12-22.

Hale, T., Held, D., & Young, K. (2013). Gridlock: Why global cooperation is failing when we need it most. Polity.

Hansen, A., & Cox, R. (2015). The Routledge Handbook of Environment and Communication. Routledge.

Harper, S.L., et al. (2021). Green Paradigms in Business: Synergies of Culture and Economics. Business & Environment, 37(2), 159-179.

Harrison, K. D. (2007). When languages die: The extinction of the world's languages and the erosion of human knowledge. Oxford University Press.

Hathaway, O. A., & Meyer, S. (2021). The consequences of economic isolationism. The Journal of Economic History, 81(1), 110-140. doi:10.1017/S0022050720000666

Hawkins, H., & Smith, T. (2020). Narrative in the Environmental Realm: Stories of Humankind and Nature. Ecocritical Perspectives, 18, 25-37.

Holling, C. S. (1973). Resilience and Stability of Ecological Systems. Annual Review of Ecology and Systematics, 4, 1–23.

Hoover, E. (2017). Natives Aren't Carbon Copies. Earth Island Journal, 32(2), 41-47.

Hulme, M. (2009). Why We Disagree About Climate Change: Understanding Controversy, Inaction and Opportunity. Cambridge University Press.

Jackson, T. (2017). Prosperity without growth: Foundations for the economy of tomorrow. Routledge.

Jenkins, W., Tucker, M. E., & Grim, J. (2018). Routledge handbook of religion and ecology. Taylor & Francis.

Jensen, R. (2021). The storytelling animal: How stories make us human. Scientific American.

Jernsletten, J. L., & Klokov, K. (2002). Sustainable Reindeer Husbandry. Arctic Council 2000-2002.

Johns, T., & Sthapit, B. R. (2004). Biocultural diversity in the sustainability of developing-country food systems. Food and Nutrition Bulletin, 25(2), 143-155.

Johnson, E., Walker, P., & Johnson, M. (2021). Sustainability education for global citizenship: The role of multinational perspectives. International Studies in Education, 22(3), 105-121.

Jorgensen, D. (1981). Kastom and Nation Building in the South Pacific. Penders.

Kates, R. W., Parris, T. M., & Leiserowitz, A. A. (2005). What is sustainable development? Goals, indicators, values, and practice. Environment: Science and Policy for Sustainable Development, 47(3), 8-21.

Katz, D. L., Meller, S., & Kavak, Y. (2014). Can we say what diet is best for health? Annual Review of Public Health, 35, 83-103.

Keck, M. E., & Sikkink, K. (1998). Activists Beyond Borders: Advocacy Networks in International Politics. Cornell University Press.

Keohane, R. O., & Nye, J. S. (2000). Globalization: What's new? What's not? (And so what?). Foreign Policy, 118, 104-119.

Kimmerer, R. W. (2015). Braiding Sweetgrass: Indigenous Wisdom, Scientific Knowledge and the Teachings of Plants. Milkweed Editions.

Kimmerer, R.W. (2013). Braiding Sweetgrass: Indigenous Wisdom, Scientific Knowledge, and the Teachings of Plants. Milkweed Editions.

Kotler, P., Lee, N., & SustainAsia Ltd. (2010). Corporate Social Responsibility: Doing the Most Good for Your Company and Your Cause. John Wiley & Sons, Inc.

Kreutzmann, H. (2016). The significance of agro-biodiversity in the Hindukush-Karakoram-Himalaya region. In: Barthel, S., Crumley, C., & Svedin, U. (Eds.), Bio-cultural refugia: Safeguarding diversity of practices for food security and biodiversity. Global Environmental Change, 41, 213-222.

Lawrence, D., & Vandecar, K. (2015). Effects of tropical deforestation on climate and agriculture. Nature Climate Change, 5(1), 27-36.

Leichenko, R., & O'Brien, K. (2008). Environmental change and globalization: double exposures. Oxford University Press, USA.

Levis, C., Costa, F. R. C., Bongers, F., Peña-Claros, M., Clement, C. R., Junqueira, A. B., & Neves, E. G. (2018). How people domesticated Amazonian forests. Frontiers in Ecology and Evolution, 5, 171.

Lu, F. (2007). Integration into the market among indigenous peoples: a cross-cultural perspective from the Ecuadorian Amazon. Current Anthropology, 48(4), 593-602.

McCarter, J., & Gavin, M. C. (2015). In situ maintenance of traditional ecological knowledge on Malekula Island, Vanuatu. Society & Natural Resources, 28(11), 1172-1189.

Mead, H. M. (2003). Tikanga Māori: Living by Māori Values. Huia Publishers.

Meadows, D. (2008). Thinking in Systems: A Primer. Chelsea Green Publishing.

Meadows, D. H. (2008). Thinking in systems: A primer. Chelsea Green Publishing.

Meadows, D. H. (2008). Thinking in Systems: A Primer. Chelsea Green Publishing.

Monteiro, M. Y., Waldstein, A., & Bourdy, G. (2011). Traditional Kuikuru Medicine in the Xingu Indigenous Park, Mato Grosso State, Brazil. Ethnobotany Research and Applications, 9, 1-27.

Mulligan, M. (2003). Feet to the ground in assessing ecological sustainability: A Maori case study. The International Handbook of Environmental Sociology, 119-137.

Nabhan, G. P. (2009). Where Our Food Comes From: Retracing Nikolay Vavilov's Quest to End Famine. Island Press.

Nicolaisen, J., & Nicolaisen, I. (1997). The Pastoral Tuareg: Ecology, Culture, and Society. Thames & Hudson Ltd.

Niinimäki, K., & Armstrong, C. M. (2013). From pleasure in use to preservation of meaningful memories: A closer look at the sustainability of clothing via longevity and attachment. International Journal of Fashion Design, Technology and Education, 6(3), 190-199.

Nyong, A., Adesina, F., & Elasha, B. O. (2007). The Value of Indigenous Knowledge in Climate Change Mitigation and Adaptation Strategies in the African Sahel. Mitigation and Adaptation Strategies for Global Change, 12(5), 787-797.

O'Brien, K., & Wolf, J. (2020). The social and community dimensions of resilience in social-ecological systems. Ecology and Society, 25(4), 2. doi:10.5751/ES-11767-250402

Orr, D. W. (1992). Ecological Literacy: Education and the Transition to a Postmodern World. Albany, NY: State University of New York Press.

Paustian, K., Lehmann, J., Ogle, S., Reay, D., Robertson, G. P., & Smith, P. (2016). Climate-smart soils. Nature, 532(7597), 49-57.

Petrini, C. (2001). Slow Food: The Case for Taste (Arts and Traditions of the Table: Perspectives on Culinary History). Columbia University Press.

Phillips, L. (2006). Food and Globalization. Annual Review of Anthropology, 35, 37-57.

Porter, M. E., & Kramer, M. R. (2011). Creating shared value. Harvard Business Review, 89(1/2), 62-77.

Posey, D. A. (1985). Indigenous management of tropical forest ecosystems: the case of the Kayapó Indians of the Brazilian Amazon. Agroforestry Systems, 3(2), 139-158.

Posey, D. A. (ed.). (2002). Cultural and spiritual values of biodiversity. United Nations Environment Programme.

Rawls, J. (1971). A Theory of Justice. Harvard University Press.

Raworth, K. (2017). Doughnut Economics: Seven Ways to Think Like a 21st-Century Economist. Chelsea Green Publishing.

Rival, L. (2002). Trekking through history: The Huaorani of Amazonian Ecuador. Columbia University Press.

Rival, L. (2002). Trekking through History: The Huaorani of Amazonian Ecuador. Columbia University Press.

Roberts, M., Norman, W., Minhinnick, N., Wihongi, D., & Kirkwood, C. (2004). Kaitiakitanga: Māori perspectives on conservation. Pacific Conservation Biology, 10(1), 7-20.

Rockström, J., Steffen, W., Noone, K., Persson, Å., Chapin, F. S., Lambin, E. F., ... & Foley, J. A. (2009). A safe operating space for humanity. Nature, 461(7263), 472-475.

Rockström, J., Steffen, W., Noone, K., Persson, Å., Chapin, III, F. S., Lambin, E. F., ... & Nykvist, B. (2009). A safe operating space for humanity. Nature, 461(7263), 472-475.

Roder, W., Dorji, K., Wangdi, K. (2001). Fallow management in Bhutanese uplands series; results from on-farm research in Chhukha District, Bhutan 1994-1997. Mountain Agriculture in the Himalayan Region, Proceedings of an International Symposium, 209-214.

Rodriguez, K., Bombay, H., & Dodo, M. (2015). The Role of Indigenous Knowledge in Desertification Monitoring and Assessment. International Journal of Biodiversity and Conservation, 7(2), 71-80.

Rodriguez, L. (2022). Sustainable fiber production in the Andes: The future of alpaca wool. International Journal of Sustainable Fashion & Textiles, 3(1), 12-26.

Schlosberg, D. (2007). Defining environmental justice: Theories, movements, and nature. Oxford University Press.

Schlosberg, D. (2007). Defining Environmental Justice: Theories, Movements, and Nature. Oxford University Press.

Schlosberg, D. (2013). Theorising environmental justice: The expanding sphere of a discourse. Environmental Politics, 22(1), 37-55.

Sen, A. (1999). Development as freedom. Oxford University Press.

Shiva, V. (1997). Biopiracy: The plunder of nature and knowledge. South End Press.

Shue, H. (2017). Climate Justice: Vulnerability and Protection. Oxford University Press.

Sidali, K. L., Kastenholz, E., & Bianchi, R. (Eds.). (2015). Food, Tourism and Regional Development: Networks, Products and Trajectories. Routledge.

Singh, A. (2001). Building a Regenerative Parara: Sukhomajri and community based watershed management. Natural Resources Forum, 25(1), 1-11.

Singh, K. D. (2013). Agroforestry: Past, present and future. In: Advances in Agroforestry. Springer, Dordrecht.

Smith, A. J., Chen, D., & Yin, J. (2019). Leveraging Traditional Knowledge to Combat Climate Change: A Framework for Innovation. International Journal of Environmental Research and Public Health, 16(18), 3355.

Smith, J., Tyszczuk, R., & Butler, R. (Eds.). (2017). Culture and climate change: Narratives. Shed.

Smith, L. T., Maxwell, T. K., Puke, H., & Temara, P. (2018). Indigenous knowledge, methodology and mayhem: What is the role of methodology in producing Indigenous insights? A discussion from Mātauranga Maori. Knowledge Cultures, 6(3), 131-156.

Sobel, D. (2004). Place-based Education: Connecting Classrooms & Communities. Great Barrington, MA: The Orion Society.

Sovacool, B. K., Hook, A., Martiskainen, M., & Brock, A. (2021). The decarbonization divide: Contextualizing landscapes of low-carbon exploitation and toxicity in Africa. Global Environmental Change, 67, 102209.

Spear, T., & Waller, R. (Eds.). (1993). Being Maasai: Ethnicity and Identity in East Africa. James Currey.

Steffen, W., Broadgate, W., Deutsch, L., Gaffney, O., & Ludwig, C. (2015). The trajectory of the Anthropocene: The Great Acceleration. The Anthropocene Review, 2(1), 81-98.

Steffen, W., Grinevald, J., Crutzen, P., & McNeill, J. (2011). The Anthropocene: conceptual and historical perspectives. Philosophical Transactions of the Royal Society A: Mathematical, Physical and Engineering Sciences, 369(1938), 842-867.

Sterling, S. (2001). Sustainable Education: Re-Visioning Learning and Change. Green Books.

Sterling, S. (2001). Sustainable Education: Re-Visioning Learning and Change. Schumacher Society.

Sullivan, M. K., Branch, L., & Hill, K. (2020). Ecuador's Waorani people 'win the first victory' in historic legal battle against the government - Rights and Resources. Rightsandresources.org. Retrieved from https://www.rightsandresources.org/

The Bhumi Project. (2020). Hindu Declaration on Climate Change.

Thiele, L. P. (2016). Sustainability (Key Concepts). Polity.

Thompson, P. B. (2015). From Field to Fork: Food Ethics for Everyone. Oxford University Press.

Thrupp, L. A. (2000). Linking Agricultural Biodiversity and Food Security: The Valuable Role of Sustainable Agriculture. International Affairs, 76(2), 265-281.

Trichopoulou, A., Martínez-González, M. A., Tong, T. Y. N., Forouhi, N. G., Khandelwal, S., Prabhakaran, D., ... & de Lorgeril, M. (2009).

Definitions and potential health benefits of the Mediterranean diet: views from experts around the world. BMC medicine, 12(1), 112.

Turner, N. J., & Clifton, H. (2009). "It's so different today": Climate change and indigenous lifeways in British Columbia, Canada. Global Environmental Change, 19(2), 180-190.

Turner, N. J., & Clifton, H. (2010). "It's so different today": Climate change and indigenous lifeways in British Columbia, Canada. Global Environmental Change, 20(2), 351-360.

UNFCCC. (2021). COP26: UN Climate Change Conference UK 2021. United Nations Framework Convention on Climate Change.

United Nations Department of Economic and Social Affairs. (2021). Sustainable Development. https://sdgs.un.org/goals

United Nations. (2001). Cultural Sustainability: The Protection of Cultural Heritage in a World in Transformation. UNESCO World Heritage Centre.

United Nations. (2007). United Nations Declaration on the Rights of Indigenous Peoples. UN General Assembly.

Verschuuren, B., Wild, R., McNeely, J., & Oviedo, G. (2010). Sacred Natural Sites: Conserving Nature and Culture. Earthscan.

Walker, B., & Cooper, M. (2011). Towards a new science of adaptation to climate change. Nature Climate Change, 1(4), 76-78.

Wangchuk, K., Wangdi, J., & Mindu. (2014). Indigenous soil and land management practices in Bhutan. Bhutan Journal of Natural Resources & Development, 1(1), 129-135.

Wenzel, G. (1991). Animal Rights, Human Rights: Ecology, Economy and Ideology in the Canadian Arctic. University of Toronto Press.

Wessels, J. (2008). Water and urban development paradigms: Towards an integration of engineering, design and management approaches. CRC Press/Balkema.

White, L. (1967). The historical roots of our ecologic crisis. Science, 155(3767), 1203-1207.

Whyte, K. P. (2013). On the role of traditional ecological knowledge as a collaborative concept: a philosophical study. Ecological Processes, 2(1), 1-12.

Whyte, K. P. (2018). Indigenous science (fiction) for the Anthropocene: Ancestral dystopias and fantasies of climate change crises. Environment and Planning E: Nature and Space, 1(1-2), 224-242.

Williams, T., & Hardison, P. (2013). Culture, law, risk and governance: contexts of traditional knowledge in climate change adaptation. Climatic Change, 120(3), 531-544.

World Commission on Environment and Development (WCED). (1987). Our common future. Oxford University Press.

Wright, T. S. (2019). The Role of Cultural Diversity in Sustainable Development. International Journal of Environmental Research and Public Health, 16(2), 262. doi:10.3390/ijerph16020262

www.ingramcontent.com/pod-product-compliance
Lightning Source LLC
Chambersburg PA
CBHW02205020426
42335CB00012B/620